ASYNCHRONOUS TRANSFER MODE NETWORKS

ASYNCHRONOUS TRANSFER MODE NETWORKS

Edited by

Yannis Viniotis

North Carolina State University
Raleigh, North Carolina

and

Raif O. Onvural

IBM
Research Triangle Park, North Carolina

SPRINGER SCIENCE+BUSINESS MEDIA, LLC

Proceedings of TRICOMM '93, held April 26-27, 1993,
in Raleigh, North Carolina

ISBN 978-0-306-44486-9 ISBN 978-1-4615-2844-9 (eBook)
DOI 10.1007/978-1-4615-2844-9

Originally published by Plenum Press, New York in 1993

PREFACE

Broadband Integrated Services Digital Network (B-ISDN) is conceived as an all-purpose digital network supporting interactive and distributive services, bursty and continuous traffic, connection-oriented and connectionless services, all in the same network. The concepts of ISDN in general and B-ISDN in particular have been evolving since CCITT adopted the first set of ISDN recommendations in 1984. Thirteen recommendations outlining the fundamental principles and initial specifications for B-ISDN were approved in 1990, with Asynchronous Transfer Mode (ATM) being the transfer mode of choice for B-ISDN.

It seems fair to say that B-ISDN concepts have changed the face of networking. The expertise we have developed for a century on telephone systems and over a number of decades on packet networks is proving to be insufficient to deploy and operate the envisioned B-ISDNs. Much more needs to be understood and satisfactorily addressed before ATM networks can become a reality.

Tricomm'93 is dedicated to ATM networks. The technical program consists of invited papers addressing a large subset of issues of practical importance in the deployment of ATM networks. This is the sixth in a series of Research Triangle Conferences on Computer Communications, which emerged through the efforts of the local chapter of IEEE Communications Society.

We would like to thank all speakers who participated in the technical program, and, Mr. Len Felton, IBM, for his keynote speech. Ms. Margaret Hudacko, Center for Communications and Signal Processing, North Carolina State University, has patiently worked with us in the organization of the conference, and we are grateful to her for making it all possible. We also would like to thank Ms. Mary Safford, our editor, and Mr. John Matzka, both at Plenum Press, for publication of the proceedings.

Yannis Viniotis
Raif O. Onvural

COMMITTEES

Conference Chairs

Yannis Viniotis (NCSU)
Raif O. Onvural (IBM, Research Triangle Park)

Program Committee

Harry G. Perros (NCSU)
Arne A. Nilsson (NCSU)
Gerald A. Marin (IBM, Research Triangle Park)
Dan Stevenson (MCNC, Research Triangle Park)
Enis Erkel (BNR, Research Triangle Park)

Local Arrangements

Margaret Hudacko (NCSU)

In cooperation with

IBM, Research Triangle Park
Center for Communications and Signal Processing, NCSU
IEEE Communications Society, Eastern NC Chapter

CONTENTS

Some Obstacles on the Road to ATM 1
D. Abensour, J. Calvignac, and L. Felton

Electropolitical Correctness and High-Speed Networking, or,
Why ATM Is Like a Nose 15
D. Stevenson

An Overview of the ATM Forum and the Traffic Management Activities.................... 21
L. Gün and G. A. Marin

Communication Subsystems for High-Speed Networks:
ATM Requirements.. 31
D. N. Serpanos

SMDS and Frame Relay: Two Different Paths Toward One Destination,
Broadband Communications.. 39
J-P. Bernoux

ATM Systems in Support of B-ISDN, Frame Relay, and SMDS Services.................. 49
P. Holzworth

Approaching B-ISDN: An Overview of ATM and DQDB.................................... 55
C. Bisdikian, B. Patel, F. Schaffa, and M. Willebeek-LeMair

On Transport Systems for ATM Networks... 75
M. Zitterbart, A.N. Tantawy, B. Stiller, and T. Braun

XTP Bucket Error Control: Enhancements and Performance Analysis....................... 89
S. Fdida and H. Santoso

Congestion Control Mechanisms for ATM Networks... 107
D. S. Holtsinger

Highly Bursty Sources and Their Admission Control in ATM Networks................. 123
K. Sohraby

A User Relief Approach to Congestion Control in ATM Networks......................... 135
I. Stavrakakis, M. Abdelaziz, and D. Hoag

Space Priority Buffer Management for ATM Networks..................................... 157
D. Tipper, S. Pappu, A. Collins, and J. George

Assignable Grade of Service Using Time Dependent Priorities:
N Classes....167
M. Perry, C. Sargor, and A. Nilsson

A Review of Video Sources in ATM Networks....179
P. F. Chimento, Jr.

Multimedia Networking Performance Requirements....187
J. D. Russell

Traffic Measurement on HIPPI Links in a Supercomputing Environment....199
H-C. Kuo, A. Nilsson, D. Winkelstein, and L. Bottomley

Index....225

SOME OBSTACLES ON THE ROAD TO ATM

Daniel Abensour, Jean Calvignac(*), and Len Felton

RTP Laboratory
IBM Corporation
Research Triangle Park, N.C.

* La Gaude Laboratory
IBM France

Abstract

The introduction of ATM will take place in the local environment first. ATM will make very high speed LANs a reality. These future LANs will be based on switches, unlike their low speed predecessors which used shared media. ATM technology will contribute to making the distinctions between LANs, MANs and WANs eventually disappear. ATM will also be the transport mechanism of choice for multimedia. However there are some alternatives to ATM. One of them is FCS which can play the role of a very high speed LAN and is positioned with respect to ATM in this paper. When it is realized, the vision of a "world wired with fibers" where the videophones have replaced the conventional telephone sets and where videoconferencing are part of the normal working environment, will draw heavily upon ATM. We are discussing some of the significant problems which remain to be solved such as the distributed support of video-conferencing. The likely rollout of ATM products and the different approaches considered to address these problems are addressed here.

Introduction

Two new magic words have recently appeared in the vocabulary associated with high speed networking: ATM (Asynchronous Transfer Mode) and multi-media. The latter refers to a new family of applications involving voice, video, still images and data. The former is one of the technologies that satisfies some of the requirements of multi-media applications, namely support of isochronous traffic. From a network standpoint, two of the key characteristics from these applications are a considerable demand in bandwidth and, in some cases, a requirement for minimal delay. Today the existing multi-media applications use *separate networks* for voice, video and

Asynchronous Transfer Mode Networks, Edited by Y. Viniotis
and R.O. Onvural, Plenum Press, New York, 1993

data and demonstrate a limited integration of functions. ATM has been conceived and produced by the telecommunications carriers to accommodate the support of voice, video and data in the same network. ATM, belonging to the packet switching technologies[1] can offer a complete granularity of bandwidth and yet accommodate isochronous traffic requirements.

It is worthwhile mentioning that when the first CCITT documents [1] were produced, ATM was only presented as *the transfer mode for implementing B-ISDN*. Since that time two things have become clearer:

- ATM has gained incredible momentum and the technology is now being "borrowed" by a wide combination of industries with computer manufacturers at their forefront. As a consequence it is now evident that ATM technology will first appear in the Customer Premises Environment (CPE).
- In the carrier world ATM is definitely making more inroads than B-ISDN (Broadband Integrated Services Digital Network). It is not clear when B-ISDN will be offered by the carriers as a collection of services while ATM based services (advertised that way) can be expected within a few years.

In this paper we describe the rationale for a roll out of ATM in the CPE. We also describe and discuss some possible alternatives. We are focusing on some of the technical problems created by the advent of ATM in the local environment, namely the coexistence of ATM switches with PBXs and the evolution of the support of classical voice functions with the introduction of video-conferences. The reader is referred to [2] for a detailed description of ATM.

Alternative Configurations in the local environment

Connectivity in machine rooms

In modern machine rooms, connectivity is changing. Both in the main-frame environment and in the super computer environment we are facing an increased and an almost "any to any" type of connectivity. In the main-frame world, the advent of a new and faster channel operating in conjunction with a dynamic non blocking switching mechanism [3] allows any CPU to be connected to any other CPU using a Channel To Channel (CTC) protocol and also allows any CPU to access any control unit thru the switch. The corresponding paths are established in a dynamic fashion and the set up of each connection is performed in a remarkably short amount of time.

In the super computer world, a similar process has taken place. The HIPPI (High Performance Parallel Interface) standard [4] includes a "switch fabric" which allows a dynamic connection of the various end systems. These systems can be heterogeneous supercomputers, archival storage devices, array DASDs, main-frame com-

---

[1] "Fast Packet Switching" is more and more often associated with ATM, even though other technologies qualify to the name.

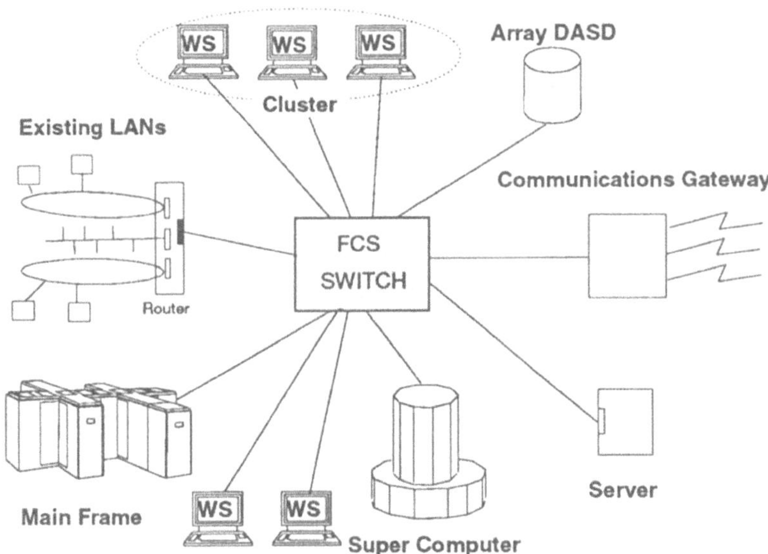

Figure 1. Potential FCS based configuration.

puters or even workstations. The utilization of a HIPPI switch results in a quite powerful but also quite expensive LAN model.

We will now see that the follow on to HIPPI, FCS (Fiber Channel Standard) [5] [6] [7] can and will play the same role, but in a much more affordable manner, and that FCS could be a real alternative to ATM in some cases. FCS was originally conceived by "channel" people as a replacement for HIPPI, IPI3 (Intelligent Peripheral Interface) and SCSI (Small Computer System Interface) types of interfaces. However, it rapidly became clear that FCS had real network attributes, at least in the local area. FCS has a wide range of speeds, up to 1.0625 Gbps or full speed, but lower speeds are also defined (half speed and quarter speed and even 1/8 speed). It covers distances[2] up to $2 \times 10Km$. FCS initially had 3 modes of operation or classes (class 1 is a point to point connection with guaranteed bandwidth, i.e. "circuit switched" type service, class 2 is connection-less service with notification of delivery or failure to deliver and class 3 is also a connection-less service with best effort delivery which is equivalent to a datagram mode). It is interesting to note that a class 4 has been recently proposed to support isochronous traffic, with the support of multi-media applications in mind.

Figure 1 illustrates a potential configuration based on FCS. Computers and workstations are directly attached to the FCS switch and communicate through the switch. As in the HIPPI case, it is expected that super computers and main-frames will also be connected using standard protocols and using paths going through the FCS switch. Again as in the HIPPI case, various devices can be directly attached to

---

[2] 10 km between the FCS switch and an attached device results in a potential "diameter" of 20 km.

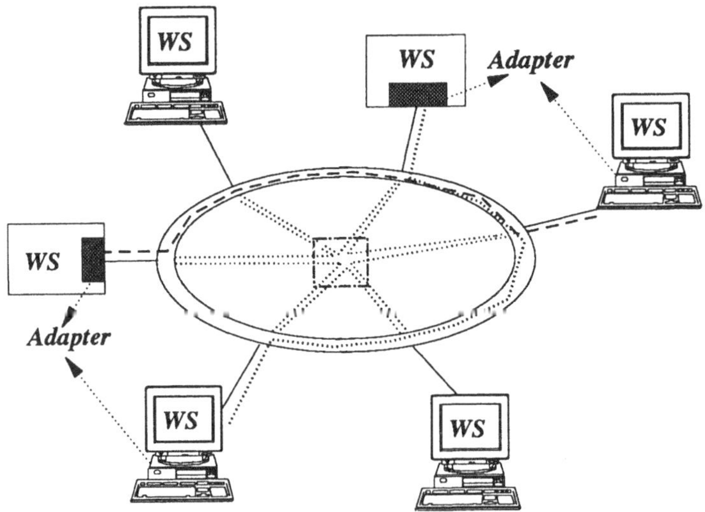

Figure 2. Shared medium based LAN

the switch. Therefore, these devices are accessible by the various processors represented. These devices may include disks arrays, frame buffers as well as print and file servers. Machines like routers may be used to allow the "old" workstations which do not have FCS adapters to communicate with the new systems and the role of the router will be to perform the required routing and protocol conversion functions. Not only will the workstations be able to communicate with each other but they can also form a *cluster*. The group of *clustered* workstations can behave like a single logical processor. This type of closely coupled processors has been developed in the past but typically each processor (or workstation) was using direct links in a point to point fashion to connect to its neighbors (using proprietary protocols). With a central FCS switch, these logical high speed links are defined through the switch and provide an equivalent fully meshed configuration with FCS connections (and using standard FC protocols). The FCS speed is a key ingredient for this type of application[3].

An important aspect of the utilization of FCS as a local network is that all the communications with the outside are funneled through a single or through multiple "gateways". Many different systems can provide the role of these communications platforms.

Switch based LAN model

We are suggesting that a switched based LAN model will be used for high speed connectivity. The classical LANs using a shared medium approach seem to be at a disadvantage when speed increases beyond 50-100 Mbps. Figure 2 depicts a clas-

---

[3] Obviously one of the key problems is the software required for the clustering (control programs and sub-systems). FCS role is to provide the necessary high speed connectivity and FCS does that, but only that.

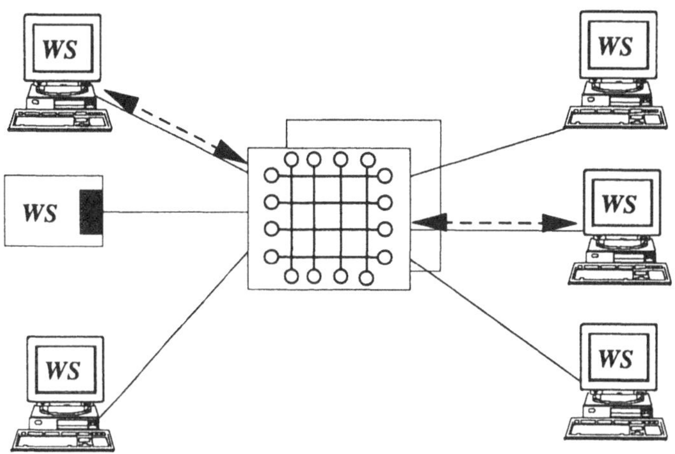

Figure 3. Switch based LAN

sical shared medium based LAN where all the attached elements (workstations, servers) compete for a common resource: the bandwidth of the shared medium. Each element has an adapter which allows the attachment to the shared medium. Each adapter must work at medium speed, even if the connection between every pair of attached elements offers only a fraction of the bandwidth of the medium (shared resource). For example if the LAN is an FDDI ring at 100 Mbps and if two workstations exchange data on the ring at, say 10 Mbps, the adapters must still work at 100 Mbps. With higher speeds, it seems that there is a significant cost difference. In addition the delay resulting from the processing by each intermediate workstation can also be significant when the two communicating nodes are separated by many of these intermediate nodes. From a topology stand point, there are no real differences between a "ring" and a switch based LAN given that rings are almost always using "wiring cabinets" and that their topology is actually more of a star than anything else. The aggregate throughput of the shared medium type of LAN has an upper limit determined by the bandwidth of the shared medium. The situation is going to be quite different with a switch based LAN where it is possible to extend the aggregate bandwidth in a virtually unlimited manner.

Figure 3 represents a model for a switch based LAN. Its major disadvantage is that it requires a switch... and the means to control the switch. This is an added cost and an added complexity. However, there are also advantages. With this approach the complexity of each end node adapter is significantly decreased since it only has to match the speed of the actual connection through the switch. Also, when a connection is established between a pair of nodes, the corresponding "bandwidth" resource is not shared. The aggregate throughput of the system can be increased by the addition of other switches in cascade or in parallel. The FCS approach follows this model and, as we shall see, so does ATM[4]. The debate comparing

[4] For ATM there is another advantage. The ports attached to the switch do not have to have the same speed. A server could use a much faster connection than a workstation. The control mechanisms will make sure that the information transfer rates are matched (leaky bucket for ATM).

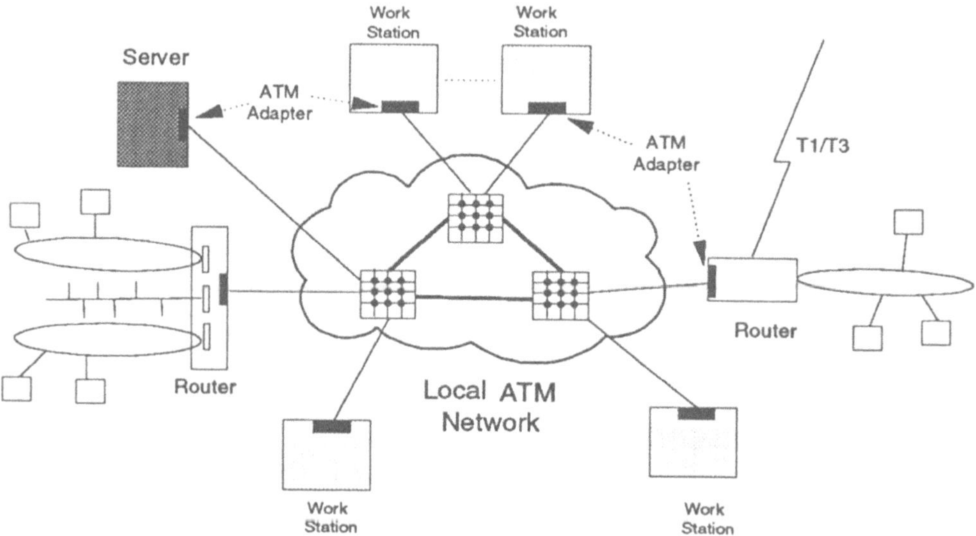

Figure 4. First phase of ATM in the local environment.

switch vs shared medium is not a new debate for LANs. There were similar discussions between PBXs proponents and LANs proponents at the time of the creation of the earlier LANs. It seems that from a technical stand point the advantages of one approach vs an other one go by cycles following the technology. But today for very high speeds and with the advent of new switching technologies and new switch architectures, all the indicators show an advantage to the switch based LAN.

ATM Rollout

Products using ATM will soon appear in the private local environment. Their introduction will be staged and the following paragraphs describe a likely scenario of the corresponding staging.

Stage 1

There are many motivations for ATM in the local environment. The key requirements that ATM address are the support of high bandwidth connectivity and the support of multi-media applications. We show a possible system configuration in Figure 4. This configuration is entirely local.

The ATM switches are systems which follow the new machine room model centered around switches [8] but are extending it to make it a new LAN. Multi-media capable workstations are attached to the ATM switches and communicate with each other through the switch. Also attached to the ATM switch are servers (like multi-media servers). Multiple workstations can have concurrent access to the server. We also find routers attached to the switch, their main purpose is to accommodate the existing LANs (Token Ring and Ethernet or even FDDI) and to allow connectivity between the existing, non ATM capable, workstations with the new ones and the new servers. The routers will have ATM adapters and will perform the necessary conversions between the existing workstations and the new ATM capable workstations.

There may be several ATM switches in the same local area and they will be interconnected by high throughput ATM links as shown in the picture. Initially speeds around 100 Mbps to 155 Mbps will be sufficient to interconnect the local ATM switches. For the workstations speeds between 25 and 45 Mbps will be satisfactory for several reasons. First a 45 Mbps link between a local ATM switch and a workstation can provide the workstation with a non shared 45 Mbps connection and there are few applications today requiring that much bandwidth. Second, the cost of the adapter is strongly influenced by the physical front end part and quite obviously a fiber interface requiring opto-electronics converters is more expensive than a lower speed front end using only copper. Third, lower speeds can be accommodated with Unshielded Twisted Pairs (UTP) at reasonable distances (at least in the local environment) and UTP saves costs of rewiring and installing new fibers. Connections to servers and inter-switch links require higher speeds because of the multiplexing of several simultaneous connections over those links.

Although this ATM model is conceptually very close to the FCS model described above, some major differences exist. ATM and FCS are intrinsically different, ATM uses a fixed 53B cell size while FCS, when operating in its "packet switching" mode, uses variable frame sizes. FCS can also operate in an almost "circuit switching" mode. More significantly, FCS has no networking background and no "communication carriers" support. Its design point was different. For example the optimization of the data transfer between disks and processors at the record level may involve very different algorithms than the ones used in the protocols associated with the transfer of data across a wide area network. ATM and the complementary architectures associated with it [9] are paying a lot of attention to the control of the network. Functions like topology service, directory service and route computation are required in a widely spread network. If FCS is indeed used as a LAN or a collection of interconnected LANs then the same functions will have to be provided[5].

What are the reasons for the utilization of ATM as a LAN in the local environment? We can consider the following:

- Natural evolution to a very high speed connectivity
- Isochronous traffic support
- Potential adapter cost advantage
- Common interface (eventually) for LANs, MANs and WANs...

High speed and isochronous traffic support are indeed functions inherent to ATM. This gives ATM a real advantage over FDDI since ATM can operate at gigabit speeds. To accommodate isochronous traffic, FDDI-II is required, but FDDI-II does not really exist. The adapter cost advantage can result from a) the significant ATM

[5] The same control mechanisms could conceivably also be adapted to, and used with FCS.

volumes and b) the relative simplicity of the adapters. The utilization of the same technology to interface a LAN or a MAN is indeed desirable but may not materialize for quite some time. The ATM interfaces on Figure 4 are all in the local environment and are not standardized by the carriers. On the other hand, the interfaces toward the public network must follow the various carriers' standards. Within the local environment, all speeds and physical interfaces will result initially from agreements between various CPE manufacturers. Currently the ATM forum is contemplating several different physical interfaces:

- 45 Mbps with DS3 framing
- 100 Mbps with 4b/5b encoding (using some of FDDI physical front end)
- 155 Mbps SONET OC3
- 155 Mbps with 8b/10b encoding (à la FCS but with different speed)

It is almost certain that this list will be extended and that some of the above interfaces will not be very successful. For example it is clear that a transmission over Unshielded Twisted Pairs (UTP category 3) at high speed rates (45 Mbps) is highly desirable and extremely likely to happen, while the intricacies of using DS3 framing and formatting (with all the wrap-tests mechanisms and the various levels of alerts) are not really required in the local environment.

In the carrier world, SONET OC3 and OC12 are the interfaces considered at this time. To accommodate the existing carrier networks it is also possible that DS3[6] ATM access be offered.

Note that the speeds considered in the local environment are not very different than FDDI. Yet it is reasonable to expect the support of very high speeds, in the gigabit range, for the applications justifying those transfer speeds, like some of the sought after FCS applications.

Stage 2

The local networks described above will be geographically dispersed and most often will need to get interconnected. Several possible approaches can be considered. Figure 5 shows the utilization of routers. These routers are using high speed private links for their interconnections. These links will normally be dedicated T1 or T3 links, but it is also conceivable to use public switched networks services for these interconnections. The choices offered today in the United States are High Speed Circuit Switched (HSCS) for switched T1 and switched T3, Frame Relay and SMDS. If the nature of the application demands isochronous traffic support then the choice is more limited and only HSCS meets the requirements, given that neither Frame Relay nor SMDS supports isochronous traffic[7].

---

[6] E3 in Europe

[7] IEEE 802.6 describes two modes of operations: pre-arbitrated and queued-arbitrated. The pre-arbitrated mode could lend itself to the support of isochronous traffic but Bellcore had elected not to support it in SMDS.

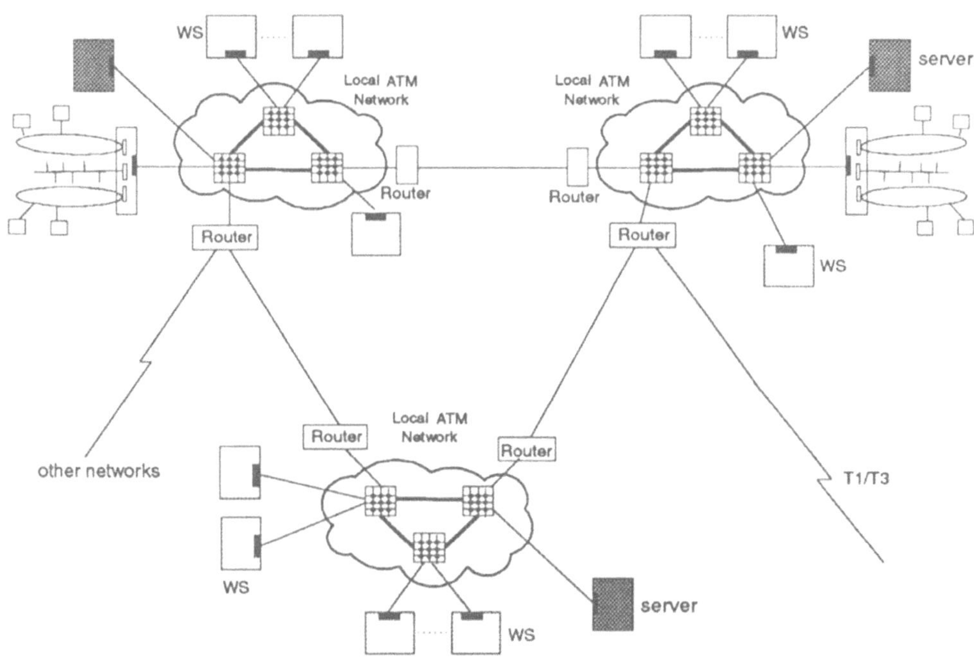

Figure 5. Interconnection of local ATM networks with routers.

For some low bandwidth applications Narrow band ISDN could also be used since it is circuit switching and, by definition, supports isochronous traffic. But then the related bandwidth would be limited to a maximum of ISDN Primary. In all cases, with this approach, the routers will have to adapt the ATM traffic to the wide area networks links used. Using routers to provide support for a wide area network has many limitations. A richer and more complete approach is shown in Figure 6

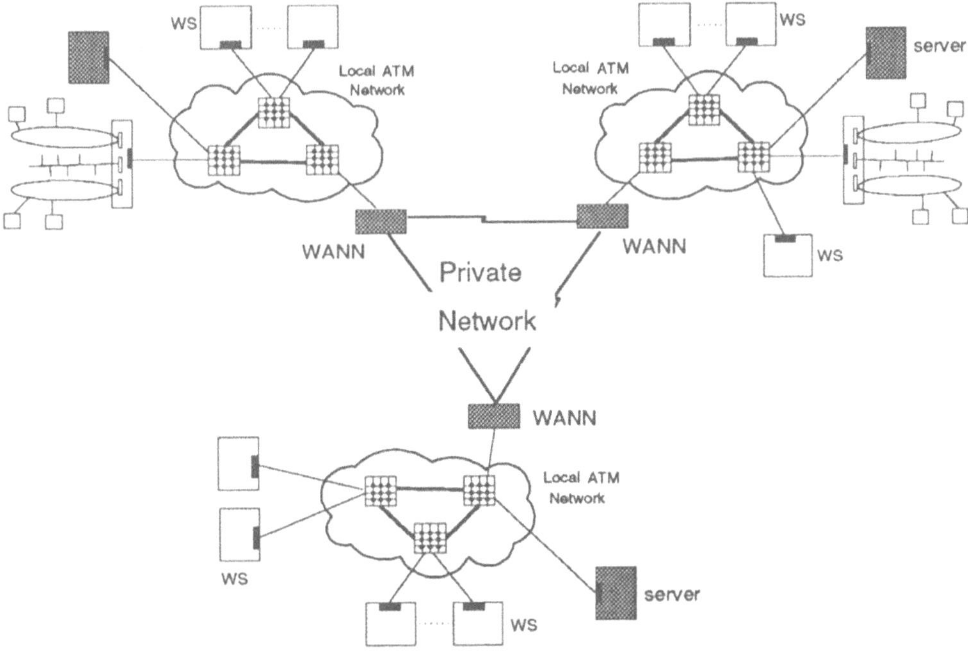

Figure 6. Interconnection of local ATM networks with bandwidth managers.

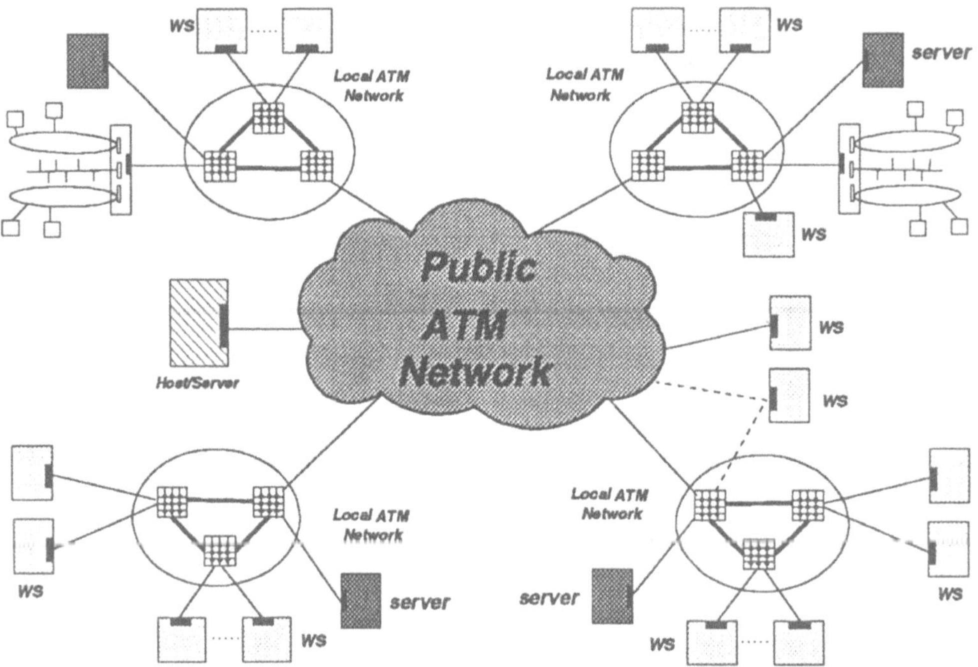

Figure 7. ATM Public switched network.

where the wide area network nodes (WANN) constitute the backbone. The nodes can use private internodal protocols or even ATM. Their function is to manage the bandwidth of the private lines, to provide the transport and to establish and control the routes. They may operate in conjunction with the routers. The key message is that a *private* backbone network will be used to interconnect the local ATM networks described in stage 1. This is likely to happen long before the advent of global ATM services provided by the carriers[8].

Stage 3

The third stage is represented on Figure 7 when a public ATM switched network is offered. We should say that this stage is not likely to happen soon since it requires active carrier involvement and a variety of services and functions which are not yet fully defined, even less standardized. These services include but are not limited to:

- directory services,
- route establishment,
- complete flow control and management control,
- Switched Virtual Circuit support,

[8] Some Virtual Private Networks based on ATM may be offered relatively soon, but they do not qualify as a global and ubiquitous service. B-ISDN does qualify but is not expected before the end of the century.

- Operations Administration Maintenance and Provisioning support,
- Billing,...

Even though this third stage is not at all ready to happen, it is still quite important since it represents the key marketing argument in favor of ATM. Figure 7 shows workstations and servers directly attached to the public ATM network. The goal is to have the same attachment valid for the local and for the wide area network, thereby eliminating some of the differences between LANs, MANs and WANs. Again there are many unsolved issues preventing this solution from happening soon. As we indicated earlier, the physical interfaces will be mostly different for the LAN and for the WAN, SONET OC3 being the only common one at this time. But physical interfaces are only a small aspect of the differences. More important and more difficult to accomplish, will be the need for common higher services: directories, route establishment, network control in general. These services are not defined in the CCITT documents, they belong in the layers above the ATM Adaptation Layer (AAL). In the private local environment, today some manufacturers are implementing their own private protocols and they will likely try to make them the base for the standards to be defined. Certainly this is going to continue to be a time consuming process.

Issues for voice and video support

Let us now look at some issues related to the support of video conferencing in this new environment. The ultimate goal is the replacement of our current telephones with devices allowing video and voice communications. These devices may initially be "multi-media capable" work stations, i.e. work stations equipped with a microphone, a speaker, a TV camera, etc., and capable of allocating one or several windows on their high resolution screens to the video conferences. Eventually we can expect to have real and relatively low cost videophones.

Very soon users may request the utilization of workstations for multi-party video conferencing and this is likely to create an interesting paradigm shift in the following manner. Today, the voice support is concentrated in an intelligent switch which can be local (PBX) or remote (CENTREX). The "end nodes" are the telephone sets, with almost no intelligence at all. The PBX typically does the call set up, the management of the outgoing lines (trunks), the collection of the billing information as well as an array of more sophisticated functions like multi-party conference, camp-on, call forwarding, consultation call, call back, system speed calling, paging, save and repeat, group pick, blocking call, etc... The software required to perform these functions is not simple and is also sizable. In addition, PBXs can prevent outside access for certain phones and can allocate different classes of services to different users.

In the ATM environment we just described, we now have a set of interconnected ATM switches and also a set of very intelligent end nodes: the powerful workstations which are going to replace our current desktops PCs. In the long range the "ATM switch" will perform all the PBX functions: control, management, administration, billing, switching as well as interfacing the outside world.

The difficult aspect of the advent of new functions is their ***staging*** relative to the existing equipment. In our current environment it is easy to have a logical connection between a workstation (PC for example) and the local PBX or more accurately with the telephone set belonging to the same user. Certain useful functions can be offered. A classical example consists in letting the user perform a directory search on the PC or on the data system to which the PC is connected; by pointing to the desired called party the user may have the dialing of the call automatically done. Also this logical connection allows the automatic display of some properties of a calling party when they are known to the PBX (the name for instance). Such functions have existed for quite some time in a local environment where modern PBXs are present.

If we now try to include video support in a local environment such as the one represented on Figure 4 and we assume that there is a local PBX (not shown on the picture), some of the control functions can be handled by the PBX. This applies to the call setup for example, or to the directory search. Switching the video signals can take place in the PBX as long as these signals are of relatively low bandwidth ($n \times 64kbps$ where $n < 7$). But since the video created by the workstation is in an ATM form, a conversion will be required (current PBXs do not understand ATM). This conversion will also be mandatory if we want the video signals to be transferred outside of the local establishment. This period can be viewed as a preliminary stage. Software changes will be required in the PBX and the resulting system will present some limitations. Again the long term solution is much simpler to conceive and to describe: a single switching mechanism (the ATM switch) and a single signalling and transport mechanism: ATM. But all the sophisticated functions performed today in the PBX will have to be migrated.

The staging of the support of video-conferencing in the ATM context will first involve some relatively simple person to person connections. Initially it will take advantage of the existing call establishment and of the other functions performed by the PBX. There will be a cohabitation and a sharing of functions between the PBX and the ATM switch. The ATM switch will be devoted to the handling of data and image. Later on the ATM switch will either include an "under the cover" PBX or will take over the PBX function. But this will come at a price: producing and testing and making work in a reliable manner the significant amount of software required.

Conclusion

We have described the advantages and some of the disadvantages of ATM for the local environment. We have also alluded to some of the real issues facing its real implementation. ATM is a promising technology and has the potential to unify many different environments. Its native support of isochronous traffic, its normal handling of very high speeds and its utilization of a switch make ATM quite attractive in the local environment. Unlike some other technologies promoted and pushed by the telecommunication carriers in the recent past, ATM stands a relatively good chance of becoming ubiquitous. However it is not realistic to expect this ubiquity to manifest itself in the next few years.

Acknowledgements

The authors would like to thank Jean Jacques Aureglia, Claude Basso and Jean Lorrain from the IBM Development Laboratory in La Gaude for their inputs and comments and they are also grateful for the reviews done by Henry Brandt from IBM Poughkeepsie, Fuyung Lai, Jerry Marin, and Gary Shippy from IBM RTP.

References

1. CCITT, Blue Book. Volume Vi, Fascicle III.4, Recommendation Q.931 (I.451), ISDN User Network Interface Layer 3 Specification, 1988.
2. J.Y. Le Boudec, "The Asynchronous Transfer Mode: A Tutorial" *IBM Research Report RZ 2133*, May 1991.
3. IBM ESCON Director Introduction. Form GA23-0363-0, September 1990.
4. American National Standard for Information Systems, High Performance Parallel Interface. X3T9.3/88-127, February 1990.
5. American National Standard for Information Systems, Fibre Channel. Physical Layer (FC-0). X3T9.3/90-0xx, November 1990.
6. American National Standard for Information Systems, Fibre Channel. Transmission Protocol (FC-1). X3T9.3/90-023, July 1990.
7. American National Standard for Information Systems, Fibre Channel. Signalling Protocol (FC-2). X3T9.3/90-019, November 1990.
8. D. Abensour, H. Meleis, A. Tantawi and D. Zumbo, Structure for High Speed Network Node RC 16421, IBM Research, January 1991.
9. I. Cidon, I. Gopal and A. Segall, "Fast Connection Establishment in High Speed Networks," *Proc. SIGCOMM'90*, pp.287-296, September 1990.

ELECTROPOLITICAL CORRECTNESS AND HIGH-SPEED NETWORKING OR WHY ATM IS LIKE A NOSE

Daniel Stevenson
MCNC, Center for Communications
Research Triangle Park, NC

ABSTRACT

With tongue planted firmly in cheek, the author of this paper summarizes the litany of popular technical complaints and objections about the suitability of Asynchronous Transfer Mode (ATM) as the means for providing high-performance communications services. The significance of these objections is discussed, and an analogy is drawn between ATM and the human nose. The point is made that many popular objections do not have much practical significance at this stage in the standardization and development of the technology. Additional problem areas are identified that the author feels are more significant, relating to deployment issues and assumptions about how ATM-based services will be used and evolve.

1.0 INTRODUCTION

The human nose, while humble in purpose, features a design that few engineers could take pride in. For instance the large sinuses in the cheekbone drain from the top. Any self-respecting engineer would have placed the drain at the bottom. The nose is runny, unsanitary and hangs upside-down over the mouth. A better design would feature a nostril at each side of the head, improving sanitation and providing a more directional sense of smell. Despite these shortcomings, few of us spend any time in debate about how to advance the state of the art in nasal design.

Odd as it may seem at first, important lessons about ATM can be drawn from the way we think about noses. Many objections can be raised about various design features of ATM. Indeed, many have been. But because the standards organizations have moved and are unlikely to undo what they have wrought, we are faced with the task of moving beyond complaint and technical argumentation to get on with the business of living with this odd thing called ATM.

2.0 THERE'S A MESS OF PROBLEMS HERE

Over the past years, many a voice has raised many an objection to a variety of technical features of ATM. Many of these objections have come from those outside the telecommunications industry who have business aspirations of developing a significant market presence there. Interestingly, in recent times, several of the founders of fast packet switching and ATM have voiced their reservations and objections to the directions being

Asynchronous Transfer Mode Networks, Edited by Y. Viniotis and R.O. Onvural, Plenum Press, New York, 1993

taken by the standards organizations [1]. Indeed, cries have gone up to turn around the process or start a new front upon which the objectors could pin their hopes for technical fulfillment. In this section, the more frequently heard complaints are identified and discussed.

2.1 Cells Are Too Damned Small

The process by which cell size was determined is both illustrative and amusing. It is also sobering to realize that engineering decisions about the analog voice network that go back to the turn of the century still have an influence on the digital broadband network. For those unfamiliar with the issues, I will momentarily digress.

At the turn of the century, telephone networks had been installed in larger cities by AT&T. There was a strong need for communications by telephone between large business centers such as Boston, Chicago, and New York, and installation of these systems was begun. Unfortunately, one of the cost-saving techniques used in local networks was not suitable for long-distance circuits. Specifically, the use of "two-wire" circuits to transmit information in both directions on one pair of wires permitted lower costs in the aerial plant for local service. For long-distance service, a "four-wire" circuit was required that provided a pair of wires each for the transmitted and the returned signals. At the interface between the local service plant and the long distance plant, a conversion between the four wire and two wire systems was needed. This was accomplished by the use of a hybrid transformer. Unfortunately, the hybrid transformers have the undesirable property of returning an echo of the signals sent toward them. In human-factors studies, it has been shown that the ear is insensitive to echoes of less than about 10 milliseconds. Echoes that return later than 10 milliseconds are perceived. It is possible to cancel echoes in both digital and analog networks, but it is an expensive function that the phone companies wish to avoid when possible.

Returning to the 20th century and ATM, the telephone companies have a tremendous capital investment in analog voice networks that will continue to be used when ATM networks are introduced. Hybrid transformers still are used in the junction between local and long-distance networks, and they still create echoes. Filling ATM cells with voice data results in a delay that varies with the size of the cell. If the cell payload were greater than 80 bytes, then this delay would be greater than 10 milliseconds, and expensive echo cancellation technology would have to be widely deployed. So the cell size was made small to avoid this problem. The choice of the 48-byte payload was a compromise between the advocates of 64-byte size and the advocates of 32-byte size.

So the bottom line is that the standards process for ATM was dominated by telephone-company interests at the time the cell size was established. The fact that the users' needs seem to been forgotten in this process is a bit of a problem. For data communications, the size of the message unit is often much larger than 48 bytes. In the arena of high-performance computing, it is generally thought that the packet size needs to increase to be able to attain the speeds supported by broadband networks. For instance, the performance of the Cray Model D input/output subsystem (IOS) is a strong function of the packet size. As part of the VISTAnet project at MCNC, we have done several experiments to benchmark the performance of this system. The high-speed channel of the Model D IOS is rated at a maximum speed of 800 Mb/s. (For those who might be curious, the low-speed channel is rated at 100 Mb/s.) The performance of the high-speed channel with 16-KB packets is about 280 Mb/s. To saturate the link requires a packet size of 256 KB. These numbers are raw I/O rates that do not take into account the impact of transport protocols in the communications path. It is thought that to run a Cray I/O system at 800 Mb/s running TCP/IP will require 1-MB packets. This corresponds to about 20,000 cells per packet.

The situation for workstations is not as extreme. The point remains valid that applications using popular network protocols that need the capabilities of broadband networks will have data units much larger than 48 bytes.

2.2 Variable-Size Packets Are Better

Within bounds, this argument seems intuitively right. In fact, the original work on fast packet switching done at Bell Labs in the early 1980s was based on the use of variable-size packets. Early in the ATM standards process, the argument about fixed-size versus variable-size packets was fought and won by the fixed-size advocates. Since then, this issue has come back to life more times than the vampire in a cheap horror movie.

In numerous forums, I've encountered engineers who have asserted that the results of their performance analyses show the superior performance of variable-size packets for certain types of applications. Frankly, I have few doubts about the correctness of their position. As noted above, high-performance applications are going to require large packets, but a large number of the packets on any data network today will be small. Screen editors need only a couple of keystrokes to require complete screen updates. The use of input devices such as mice create large numbers of small packets. There is obvious value in the ability to efficiently accommodate the needs of high-performance applications, voice traffic, and mouse movement.

2.3 The Standards Aren't Complete

There are two substantive areas yet to be resolved in the area of ATM standards: signaling and bandwidth management. The general outlines of signaling are understood. A variation of Q.931 will be used. There are numerous details that must be reconciled. A draft recommendation is scheduled to be completed by CCITT in 1993. The situation for bandwidth management is more difficult. The processes of policing and bandwidth allocation are both still research topics. Only the simplest of techniques (policing and allocating for the peak bandwidth) is thought to be understood and workable. The popular leaky bucket technique has intuitive appeal, but careful analysis and network simulations have raised serious doubts about its viability [2,3,4,5]. Allocating for the peak bandwidth is undesirable because of the obvious inefficiencies of bandwidth use. Although this may be acceptable in a LAN environment (where bandwidth is inexpensive), for WANs, the economic inefficiency associated with this approach will be a major problem. It is likely that early WANs necessarily will suffer this inefficiency and one of the significant advantages of ATM will not be realized initially. It seems likely that by the time ATM WANs begin to have a significant penetration in the physical plant of the local and inter-exchange carriers, practical solutions will be found for the problem of bandwidth allocation and policing. Admittedly, this is a statement of faith.

2.4 Adaptation is Complex

The specifications of AAL3/4 bring to mind the old joke about camels and committees. The process as defined is sufficiently complex that it is hard to imagine how it is going to ever be used for high-performance applications that have a need for broadband services. Hardware-based adaptation seems cumbersome. However, there are vendors who have quietly if not openly talked about adaptation chips that implement AAL3/4. Software-based implementations are not likely to run particularly fast. Our own experience at MCNC supports this expectation.

There is some hope from the AAL5 approach. It is simple enough to expect efficient software implementations. At MCNC, we have designed AAL5 hardware that can sustain rates of 622 Mb/s, scalable to 2.5 Gb/s using off-the-shelf components. Software implementations of AAL5 are pending.

One of the problem areas associated with adaptation is not the complexity of the process but the difficulty of dealing with large packets on high-speed links in the reassembly process. The problem becomes quite evident if you think about the interface between a high-speed (non-ATM) LAN and a ATM network. Such an interface (which we should think of as an ATM gateway) is required to manage the fragmentation and reassembly of packets for a potentially large number of active virtual connections or paths. Because some of this traffic may involve the use of large packets, the memory required for reassembly may become large. Cells arriving from the ATM side of the gateway come from random sources and are associated with random virtual connections or paths. They must be reassembled into complete packets before delivery to the LAN. Potentially complex memory management schemes must be involved. For example, at MCNC we are designing an experimental gateway that will interconnect HIPPI networks to ATM networks for supercomputer applications. The system uses a page-oriented memory management scheme for reassembly and uses a reassembly buffer of 64-MB capacity to support a maximum of 32 simultaneously active virtual connections.

3.0 NOT!

The harsh fact is that all the above objections matter little. The ATM cell is fixed in size and that is not going to change. The standards for AAL3/4 will not likely change, although AAL5 may become more widely used. The lack of complete standards has not slowed down the introduction of ATM-based LANs by a variety of vendors. There may be many aspects of ATM that we do not like, but it works. Furthermore, the telecommunications industry will grind along implementing ATM systems as the building blocks for broadband networks. They are not likely to embrace alternative technologies for a long time to come. Remember the nose. Do we rant and rave about how poor the design is? Of course not! We get on with life. We invent tissues and handkerchiefs. We socialize our children to blow their noses. For the most part we live out or lives with out paying much attention to the messy, disgusting, inefficient features of the nose and go on with it. We are best served by adopting this practical approach for ATM as well, not dwelling on the inconveniences, and learning how take advantage of it capabilities it offers.

4.0 PROBLEMS THAT WILL BITE OUR BACKSIDES

4.1 Telcos Don't Understand Data

The first 13 years of my career involved working in the R&D subsidiaries of several telecommunications vendors. My experience for those years was that telcos have a dominant culture that centers around circuit-based voice communications. Despite the excellent talent that most of them have involved in data communications technologies, products, and services, my observation has been that ultimately they all think and act in ways that center around voice- and circuit-oriented communications. Some of this behavior is not entirely of their own doing.

I can cite some examples of this kind of bounded thinking that perhaps bring out the points that need to be made here. In the VISTAnet[1] project, we have several high-end computing resources that are interconnected by an ATM network. The interface between the network and the host is a system component known as the Network Terminal Adapter (NTA). Its function is multifold, but the primary purpose of the NTA is to convert between the 800-Mb/s HIPPI channel [6,7,8] of the hosts to ATM cells on synchronous optical network (SONET) OC-12c local loops. There will be four hosts attached to the VISTAnet ATM network, and there is an NTA for each. The functionality of the NTA was the subject of some argument at the beginning of the project. Since this unit is the design responsibility of Fujitsu and BellSouth, they have opted for functionality that their engineers feel comfortable with. As a result, the NTA functions as a HIPPI extension device with limited multiplexing capability; that is, HIPPI packets from the host are converted into cells sent across the ATM network and then reconverted into HIPPI packets. The link-layer protocols associated with HIPPI are not terminated by the NTA. It assumes and requires that it talks with another NTA on the other side of the ATM network. A workstation directly attached to the same ATM network cannot communicate with an NTA-attached host.

A similar situation exists with plans that are being formulated by GTE, Carolina Telephone and BellSouth for the early introduction of ATM capabilities in the state of North Carolina. The network is planned to go into service late 1993 providing OC-3c rate ATM connectivity to about 60 cites across the state. Although it will be possible to directly connect terminal devices to the network few vendors will support the required interfaces in this time frame. To deal with this general lack of interfaces the plans call for the use of a system element known as the service multiplexer. The service multiplexer supports a variety of interfaces so that different services can be provided to end users. The types of services include delivery of T3 channels, NTSC channels, and point to point. It should be clear from the description that the service multiplexer provides the means for carving point-

---

[1] The VISTAnet project is supported by BellSouth, by GTE Corp. and by the National Science Foundation and the Defense Advanced Research Projects Agency under cooperative agreement NCR 8919038 with the Corporation for National Research Initiatives.

to-point circuit-oriented services off of the ATM network. This approach has been compared to that of a farmer who upon buying his first tractor hitches his mule to the front of it to plow his fields [9]. The value for data communications of ATM (versus SONET alone) is its ability to support multiplexed communications, not just over the telco facilities but to deliver them all the way to the terminal devices. This ability simply is not captured in the plans being formulated.

4.2 Fixation With 155 Mb/s

Again and again when talking with telcos, I find them uninterested in the problems associated with applications that require communications running faster than 155 Mb/s. They are consistently reluctant to believe that there will be any significant future demand for services above 155 Mb/s. Pointing out the existence of such applications today in the supercomputing environment does not change this impression. Supercomputing users represent a small community that telcos seem to feel they can afford to ignore.

The problem with getting attached to one data rate as the answer to all data communications needs is the rapid rate of change in the computer industry. In the computer industry, a new generation of technology emerges every three to five years. We are all familiar with the rapid pace of change in the areas of memory density and processor speeds. Amdahl's law [2] leads us to expect that the speed requirements for data communications will increase exponentially along with the processor speeds and memory density. This implies that the data rates that look more than adequate for today's applications eventually will become inadequate. Unfortunately, the amount of time required for standards development and acceptance and system deployment in the telecommunications industry is several times longer than one generation of technology in the computer industry. This scenario was played out in the development of basic-rate ISDN technology.

The requirements of today's supercomputers will be the same as that of tomorrow's desktop workstations. As recently as five years ago, it was commonly thought that only supercomputing centers had the ability and need to drive long-distance communications networks at T1 (1.544-Mb/s) data rates. Today, T1 and T3 rate networks are common.

4.3 Striping as a Path to Faster Service

There are two basic ways to support applications needing higher speeds than 155 Mb/s: the concatenated approach and the striped approach. The concatenated approach requires making switches, host interfaces, and other network infrastructure that can deal with 622 Mb/s and higher channels as a single pipe of information. Conversely, the striped approach requires making switches, host interfaces, and other network infrastructure that can deal with multiple 155 Mb/s channels multiplexed into a faster carrier. The striped approach has a natural appeal to the phone companies. It means that they can concentrate exclusively on 155 Mb/s infrastructure, and the folks that need more bandwidth do not require any special effort. And since there aren't many of them around, why should special provisions be made for them?

There are two fallacies in this argument. The first is the short-sightedness of this vision. As I have argued earlier, in the world of data communications, today's lunatic fringe is tomorrow's bread and butter. The second problem is that some applications are going to remain sensitive to cell ordering, and preserving cell ordering for ATM across STS-3c boundaries is difficult. As a consequence, the efforts to build high-speed interfaces using striped ATM [5, 6] take the design approach of passing each datagram (comprised of many cells) across a dedicated stripe; that is, the first datagram goes over stripe 1, the second goes over stripe 2, and so on. Each individual datagram has available to it only 155 Mb/s of bandwidth. For circumstances in which datagrams are large, the transmission delays are lengthened by forcing each datagram to pass through the 155 Mb/s stripe. Overall system throughput can suffer if the staging buffer is not sufficiently large.

---

[2] Amdhal's law states that a well balanced computer has 1 byte of memory and 1 bit per second of I/O for every instruction per second.

5.0 CONCLUSION

In this paper, I've indulged in creative ranting to reveal the absurdity of popular electropolitical debates about ATM and its viability as a technology for high-performance communications. The point has been made that if we can get along with a design as silly and impractical as the human nose, we can do so also with ATM. Our ability as individuals or corporations to change the basic aspects of ATM are just as limited as our ability to modify the way our noses are implemented. Nonetheless, there are some important aspects of how ATM is being deployed and used that can result in serious future problems for service providers. Focusing too exclusively on the needs of customers today can lead to getting caught short in an era of rapidly escalating need for communications speeds.

6.0 REFERENCES

[1] Private communications with M. Decina and D. Vlack.

[2] D. Holtsinger and H.G. Perros, "Performance analysis of leaky bucket policing mechanisms". Submitted to IEEE Trans. Comm.

[3] D. Holtsinger and H.G. Perros, "Waiting time and cell loss probability analysis for the buffered leaky bucket". Submitted to a special issue of the J. Performance Evaluation.

[4] D. Holtsinger, "Design and analysis of the dual leaky bucket policing mechanism for ATM networks". Submitted to ICC '93.

[5] D. Holtsinger, "Performance analysis of leaky bucket policing mechanisms for high-speed networks". Ph.D. thesis, NC State University. November '92.

[6] American National Standard for Information Systems. "HIPPI Mechanical, Electrical, and Signalling Protocol Specifications (HiPPI-PH)", X3T9.3/88-023 Rev 7.2. August 2,1990.

[7] American National Standard for Information Systems. "HIPPI Framing Protocol Specifications (HiPPI-FP)", X3T9.3/88-023 Rev 2.6. July 24,1990.

[8] American National Standard for Information Systems. "HIPPI Link Encapsulation of ISO 8802-2 (IEEE std 802.2) Logical Link Control Procol Data Units (HiPPI-LE)", X3T9.3/90-119 Rev 2.0. December 3,1990.

[9] Private communication with Dan Winkelstein.

AN OVERVIEW OF THE ATM FORUM AND THE TRAFFIC MANAGEMENT ACTIVITIES

Levent Gün and Gerald A. Marin

IBM, Network Analysis Center
Networking Systems Architecture
Research Triangle Park, NC, 27709, U.S.A.

ABSTRACT

In this paper we provide an overview of the ATM Forum objectives and summarize the work in progress in the Traffic Management Subworking group as of year end 1992. We give particular emphasis on traffic characterization since this is one major area where the ATM Forum has taken decisions beyond that of the CCITT Standards.

1.0 ATM FORUM OBJECTIVES

Asynchronous Transfer Mode (ATM) is a cell relay technology for integrated networking of data, voice and video at transmission speeds of several megabits per second to multiple gigabits per second. It is the basis for Broadband Integrated Services Digital Network (B-ISDN). Since its initial recommendation in 1988 by CCITT ATM has quickly become a popular technology in the communications and computer industries. Due to the progress that has been made in the last 5 years ATM is no longer viewed as a technology in the distant future, but as a technology that can be deployed internationally within a few years.

The ATM Forum is an international consortium chartered to accelerate the use of ATM products and services through a rapid convergence and demonstration of interoperability specifications, and promotion of industry cooperation and awareness. Since its conception in October 1991, the member list has grown

Asynchronous Transfer Mode Networks, Edited by Y. Viniotis
and R.O. Onvural, Plenum Press, New York, 1993

rapidly to over 160 organizations and continues to grow (see Appendix). Industry sectors represented include LAN and WAN vendors, internetworking vendors, computer vendors, switch vendors, local and long distance carriers, PTTs, semiconductor manufacturers, government agencies, research organizations and independent consultants.

The ATM Forum is not a standards body, but instead works in cooperation with standards bodies such as ANSI and CCITT. The areas of focus so far have been the physical layer, the ATM layer, the ATM Adaptation Layers, Signaling, and Traffic Management. In June 1992 the ATM Forum published a User Network Interface (UNI) Specification document [1] that addresses the Physical Layer, ATM Layer, and the Adaptation Layer agreements reached in the first six months. The technical Committee is now focused on writing a signaling and a traffic management document. Drafts of these documents are to be completed by sometime in Spring'93.

There are two principle working committees in the ATM Forum: the Technical Committee and the Market Awareness and Education Committee. The Market Awareness and Education Committee meets 12 times a year to discuss issues relative to the goals of the ATM Forum, the need to offer ATM Education (and ATM Forum education), the creation of documents and presentations describing the work of the Forum, and other topics related to positioning of ATM in the marketplace. There are four subworking groups now including the End Use Focus Group, the Education Group, the Marketing Group, and the Strategic Planning Group. The Education Group has a responsibility to generate educational materials such as those showing how Frame Relay, SMDS and ATM complement each other. Market Awareness is responsible for preparing and arranging presentations of ATM and Forum overview materials. The End User Group focusses on determining end user needs and concerns. The Strategic Planning Group concentrates on directional issues such as strategies for internetworking with other technologies and definition of evolutionary paths from currently installed technologies to ATM.

The ATM Forum's Technical Committee also meets 12 times per year and has an informal review and approval cycle compared with other standards bodies. Currently work is progressing in two main subworking groups: Signalling and Traffic Management. In addition there are two other parallel working groups looking at more specialized interfaces: Inter-Carrier-Interface (ICI) and Data Exchange Interface (DXI).

The Signalling subworking group is currently focussing its efforts into agreeing on the signalling requirements for Phase 1. The current intend is to use the Q.93B specification of CCITT as the base with minimal extensions for Phase 1. Other topics of discussion are reference configurations, addressing and address resolution, end-to-end compatibility and point to multipoint signalling.

The DXI group is working on an interface to attach existing Data Terminal Equipment such as routers that support traditional "frame" interfaces (such as V.35) to ATM networks. The ICI group is working on inter carrier service description for public ATM network carrier-to-carrier connections. Initial ICI specification will support permanent virtual circuit (PVC) based ATM cell relay service, PVC based ATM frame relay service and SMDS. The ICI will be capable of supporting multiple services.

In this paper, we only focus on the activities of the traffic management sub-working group and provide an overview of the agreements to date as well as work currently in progress.

2.0 TRAFFIC MANAGEMENT ACTIVITIES

Traffic Management subworking group currently works on drafts of the traffic management baseline text and Quality of Service (QOS) text.

The scope of the initial work on Traffic Management is limited to a set of Traffic Management functions and procedures necessary for the completion and delivery of the next release of the UNI specification. Therefore, the group is focussing initially on a restricted set of congestion control capabilities using simple mechanisms to achieve adequate network efficiency. Additional traffic control mechanisms will be considered in later phases to achieve increased network efficiency.

The general approach taken is to use the CCITT I.371 document [2] as a reference and to concentrate initially on traffic management issues involved at the public and private UNIs. To date the focus of the group has been on the user-network traffic contract definition. In particular, the traffic parameter specification. This is one area the ATM Forum has made decisions beyond CCITT by defining traffic descriptors in addition to the peak cell rate. We shall discuss this in detail in Section 2.2. The work on the traffic control functions and procedures is just starting and the initial draft text is likely to follow that of CCITT recommendation I.371. This includes Connection Admission Control (CAC), Usage Parameter Control (UPC), cell shaping, priority control and selective cell discard. Similarly, work on congestion control functions and procedures at the UNI is in its infancy and is likely to follow the text in I.371. This includes explicit forward and backward congestion indication and selective cell discard.

Traffic parameters describe traffic characteristics of an ATM connection. In addition to descriptors defining the traffic parameters, the requested QOS class, the cell delay variation and burst tolerance, and configuration rule for traffic conformance (see Section 2.2) define the traffic contract. The CAC procedures then determine if sufficient network resources exist to accept the connection. Once the connection has been accepted, the QOS requested is provided for all traffic that is compliant with the traffic contract. Testing compliance with the traffic contract is the responsibility of the UPC actions. UPC algorithms monitor and test the conformance of the traffic stream to the negotiated traffic parameters. If many cells of a connection is non-conforming with the traffic definition, the network may declare the connection as non-compliant. The definition of non-compliant connection is network specific. However, it is included in the traffic contract so that the user knows the exact definition. For non-compliant connections, the network need not respect the agreed QOS. However, for compliant connections, it is the network's responsibility to provide the requested QOS for all cells that are compliant with the traffic definition. The present thinking of the traffic management group is that the UPC and CAC procedures are operator specific.

2.1. Quality of Service

The ATM Layer Quality of Service (QOS) is defined by a set of parameters such as delay and delay variation sensitivity, cell loss ratio, etc. This QOS is part of the Traffic Contract at connection establishment.

There are two classes of QOS definitions, the ***specified class*** and the ***unspecified class***. The specified class contains a finite number of vectors of QOS parameter values supported by the network. This QOS class provides a QOS service to an ATM virtual connection in terms of a set of specified objective values for a subset of the ATM performance parameters (which are not yet fully defined). The general idea is that if the user specifies at least one QOS parameter, its request will be mapped to one of these pre-defined classes by the network. The unspecified class contains a single QOS vector with all of its entries null, i.e., the traffic contract contains no QOS parameters. However, the network provider may choose to map this unspecified class into one of the QOS classes supported by the network.

The Cell Loss Priority (CLP) bit of the ATM cell header allows for two Cell Loss Ratio objectives for a given ATM connection. The cells that are marked as CLP=1 are more likely to be dropped during transient congestion at the network switches than the unmarked CLP=0 cells. Therefore, a user may request two different cell loss ratio objectives for a single ATM connection. When the user requests two cell loss ratios, one is for the CLP=0 traffic and the other is on the CLP=1 traffic. Note that this is a deviation from CCITT's position (see [2]) where the cell loss ratios are specified for the CLP=0 stream and the aggregate CLP=0+1 stream. One reason for this deviation is that, in general, the cell loss ratio guarantees for the two streams can have qualitatively different meanings. For example, the cell loss ratio for CLP=1 stream may have a "looser" definition, e.g., it may represent fraction of CLP=1 cells lost over an aggregate of connections as well as over a longer time scale than that of CLP=0 cells. Furthermore, defining cell loss ratio objectives for the two streams separetely is more meaningful both for the end user applications such as layer coded video as well as for the network in designing cell discard policies.

2.2. Traffic Parameters

In order to have an operational definition of the traffic parameters the traffic management group agreed that the traffic parameters of an ATM connection be defined with respect to a deterministic rule. The main advantage of such a rule based definition is that it is easy to determine, both by the user and the network, whether a cell is compliant with the definition or not. After lengthy discussions on what the rule defining the traffic parameters should be, it is decided that the traffic definitions be based on the "leaky bucket" algorithm.

The operation of the basic leaky bucket control mechanism is as follows: credits are generated to a credit pool of size M at some fixed rate R Each credit gives permission for transmission of a cell, i.e., R is in cells/sec and M is in cells. A cell can enter the network only if there is credit available. After each cell sent into the network the number of available credits is decremented by one.

In brief, traffic parameters of a cell stream are defined in terms of two leaky bucket based traffic descriptors: The peak cell rate descriptor and the sustainable cell rate descriptor. Each descriptor is defined in terms of a continuous-state version of the discrete state leaky bucket mechanism defined above. This version of the leaky bucket algorithm, also proposed by CCITT in Annex I of I.371 (see [2]), is used as a generic cell rate algorithm, and is called $GCRA(T, \tau)$, where its parameters T and τ are related to R and M by

$$T = 1/R \text{ and } \tau = (M-1)/R .$$

Note that both T and τ are in units of time. The parameter τ can be viewed as the amount of time variation the leaky bucket will allow a cell from its theoretical arrival time, which is at equally spaced intervals of length T (see [2] for details).

Three traffic parameters are identified to partially characterize the cell generation stream at the source. These are:

1. peak cell rate R_p,
2. sustainable cell rate R_s, and
3. maximum compliant burst size B_c.

They are defined using two GCRA (T, τ) descriptors; GCRA $(T_p, 0)$ and GCRA (T_s, τ_s), where

$$T_p = R_p^{-1} \ , T_s = R_s^{-1} \text{ and } \tau_s = (B_c - 1)(T_s - T_p) \, .$$

A user may be attached to a public-UNI either directly or through a customer premise network. In either case, the user cell stream incurs delay variation caused either by traffic shaping at the customer premise network and/or by multiplexing cells of multiple connections on to the physical access channel, and due to the insertion of Operation and Maintenance cells. It is the responsibility of the user to account for this delay variation between the point of cell generation and the UNI. Therefore, the user "adds" (not necessarily linearly) the effects of these delay variations to its original traffic descriptors GCRA $(T_p, 0)$ and GCRA (T_s, τ_s), to characterize its traffic at the UNI through the descriptors GCRA (T_p, τ) and GCRA (T_s, τ'_s), where τ and τ'_s, $\tau'_s > \tau_s$, account for the cell delay variation. In fact, the parameter τ is called the cell delay variation (CDV) tolerance, while the parameter τ'_s is called the burst tolerance. Therefore, as far as the traffic contract is concerned, the effect of multiplexing on the original cell stream parameters are summarized by four parameters; the peak cell rate, the sustainable cell rate, the CDV tolerance and the burst tolerance. These four parameters, defined through GCRA (T_p, τ) and GCRA (T_s, τ'_s) are part of a traffic contract.

The peak cell rate specifies an upper bound on the traffic that can be submitted on an ATM connection. The peak cell rate and the CDV tolerance are mandatory parameters and are supplied either explicitly or implicitly during the set-up. The other two traffic parameters, the sustainable rate and the burst tolerance are optional parameters. They allow a finer definition of the traffic characteristics that enable the network to do more efficient resource allocation.

The reader should note that while the rule based deterministic traffic descriptors are attractive at the public-UNI where a "legal" contract may exist between the public network and the user, it comes at a price, namely, loss of efficiency. Since the traffic is defined in terms of deterministic rules, it does not carry the necessary information to characterize the statistical behavior of the stochastic cell arrival process. Since the "legal" value of such determinism is far less in a private-UNI, it is possible that the private-UNI may not support the same traffic contract as does the public UNI.

In general the traffic description of an ATM connection may be more complex if the connection contains multiple cell streams, i.e., CLP=0 and CLP=1 streams. If a user requests two levels of priority for an ATM connection, as indicated by the CLP bit value, the intrinsic traffic characteristics of both cell flow components have to be characterized in the source traffic descriptor through a GCRA based configuration rule. This is by means of a set of traffic parameters associated with each component. In general, how different GCRA algorithms for different descriptors interact with each other depends on a particular application.

Therefore, the group did not pick a particular configuration but instead provided example configurations (see [3]). For example, the traffic parameters of CLP=0 and CLP=1 streams can be defined by a rule determined by three GCRA algorithms. One defining the peak rate of the aggregate CLP=0+1 stream, one defining the sustainable rate and maximum compliant burst size of CLP=0 stream and a third defining the sustainable rate and maximum compliant burst size of CLP=1 stream. The exact rule can be defined as follows: If a cell fails the peak rate leaky bucket it is declared non-conforming. A CLP=0 cell that is compliant with the peak rate descriptor is checked against the sustainable cell rate descriptor for CLP=0 stream. If it is also compliant with this descriptor than the cell is compliant. If it fails the second test, it may be tagged as CLP=1 and tested against the sustainable cell rate descriptor for the CLP=1 descriptor. Similarly, a CLP=1 cell that conforms to the peak rate descriptor is tested against the sustainable cell rate descriptor for the CLP=1 stream. If it is also compliant with this descriptor, then the cell is compliant. Otherwise, the cell is declared non-conforming. Note that this algorithm only determines whether or not a cell is conforming to the rule based traffic definition. It is the function of the UPC algorithm to determine the appropriate action on the non-conforming cells. For example, the UPC algorithm may tolerate a given fraction of non-conforming cells before it takes discard actions.

Finally, we note that while traffic shaping by the user is an optional function, it is recommended that the user shapes its traffic by an algorithm that mimics the rule that defines the traffic descriptor. With such a shaper in place, the user can convert a non-compliant cell stream to a compliant one by reshaping its traffic at the cost of some scheduling delay.

3.0 SUMMARY

In this paper we have provided an overview of the work in progress in the Traffic Management Subworking Group of the ATM Forum. While the general approach is to stay in line with the CCITT standards as much as possible so far the Forum has moved beyond what is defined in the CCITT in some areas. One such area is the definition of the traffic parameters in addition to peak cell rate. In this overview we have provided these additional traffic definitions and have tried to provide a flavor of work in progress on traffic management issues in the ATM Forum.

REFERENCES

[1] "ATM User-Network Interface Specification", Version 2.0, ATM Forum, June, 1992.

[2] CCITT Draft Recommendation I.371, "Traffic Control and Congestion Control in B-ISDN", June, 1992.

[3] L. Gün and D. Hsing, "Examples of Rules Specified in a Traffic Contract", ATM Forum Contribution 248, Dec, 1992.

APPENDIX

Principal Members to the ATM Forum:

3Com, ADAPTIVE Corporation, ADC Kentrox, Alcatel, Alliance Consulting, Inc. Ameritech Services, Apple Computer, ascom Timeplex, AT&T, BBN Communications Bear-Stearns and Co, Bell Atlantic, Bellcore, BellSouth, Cabletron Systems, Cisco Systems, CNT Corp, Digital Equipment Corporation, Digital Link, DSC Communications Corporation, Ericsson, Fore Systems, Inc., France Telecom, Fujitsu, GTE Government Systems, Hewlett-Packard, Hitachi Telecom USA, Hughes LAN Systems, IBM, MCI Communications, Motorola, Inc., MultiMedia Networks, National Semiconductor, NEC America, Netrix, Network General Corporation, Network Systems, Newbridge Networks, Nokia, Northern Telecom, NYNEX, Pacific Bell, QPSX, Retix, Rolm Systems, Southwestern Bell, Sprint, Stratacom, Sun Microsystems, Synernetics, SynOptics, Telematics International, Ltd., Texas Instruments, Thomson-CSF, U S West, Ungermann-Bass, Wellfleet Communications, WilTel.

Auditing Members to ATM Forum:

Advanced Micro Devices, Advanced Network & Services, Allied Telesis, Inc., Amoco Corporation, AOTC, Artel Communications, ascom US Tech, Base2 Systems, Bell Canada, British Telecom, Bytex Corporation, CascadeCorporation, CERN, CHI Systems, Inc., Chipcom Corporations, Compression Labs, Coral Network Corporation Cray Communication, Crescendo Communications, CRS4, CSELT, Cypress Semiconductor, Department of Defense, Dynatech Communications, E-Systems, Inc. EDS, Emulex Corporation, FiberCom, Inc., Fibermux Corporation, Force Computers, Fujikura Technology America, Fujitsu Microelectronics, Ltd,.Gandalf Systems Corporation, Gartner Group, General DataComm, Inc., Grandsen Group, GTE Labs, GTE Telephone Operations, Info/Mation, Infonet Services Corp, InteCom, Intel, Intel (Israel), Interphase, Kendall Square Research, Lawrence Berkeley Labs, Madge Networks, McQuillan Consulting, MFS Datanet, Inc., MicroUnity Systems Eng., Mitel, Monarch Information Networks, Motorola Codex, MPR Teltech Ltd., N.E.T., NCR, NEC Electronics, Inc., Netcomm Ltd., NetExpress, Inc., Netvantage NeXT, Novell, NTT, OKI America, Inc., Omnitele, Optical Data Systems, Philips Kommunikations, Proteon Inc., Protocol Engines, RTFC, Inc., SAIC, Siemens Stromberg-Carlson, Silicon Graphics, Inc., Silicon Systems, SIP, Standard Microsystems, Starlight Networks, Inc., Sumitomo Electric U.S.A., Swedish Telecom, T3plus Networking, Inc., Tandem Computers, Inc, Tekelec, Tektronix,

Telco Systems, Telecom Finland, Telecommunications Labs, Teleos Communications, Themis Computer, TRA, Transwitch, TRW, Ultra Network Technologies, V-Band Corporation, Verilink, VLSI Technology, Wandel and Goltermann, Washington University, Xerox, Parc, XLNT Designs, Inc., Xyplex, Inc.

COMMUNICATION SUBSYSTEMS FOR HIGH SPEED NETWORKS: ATM REQUIREMENTS

Dimitrios N. Serpanos
IBM, T. J. Watson Research Center
P.O. Box 704, H2-D16
Yorktown Heights, NY 10598

Abstract

ATM traffic places strict performance requirements on communication subsystems. Conventional subsystems attaching to networks that use variable size packets provide good perorfmance for long packets but they are limited for short packet traffic, such as fixed cell ATM traffic. In this paper, we present the necessary changes to a conventional adapter in order to support ATM attachments. We describe the support required for packet segmentation and reassembly, and we introduce the design of a network attachment that employs a buffer cell admission policy to increase the number of successfully received packets. A distributed architecture for data memory management is also described. Adoption of the presented architectural enhancements can greatly improve performance as simulation results indicate.

1 Introduction

The dramatic progress in transmission link technology during the last decade has brought new challenges in the architecture of communication subsystems which provide connectivity among networks in both the local and the wide area. Conventional systems offer connectivity not only to traditional networks, such as Ethernet [3] and Token Ring [1], but to high speed networks, such as FDDI [4], and wide area links as well.

Communication subsystems are typically organized as a set of adapters intercommunicating over a backbone interconnection such as a bus or a switch. The adapters attach to specific networks providing storage for data packets and processing power for executing the appropriate protocols. The increasing speed of employed transmission links requires new aggressive designs for adapters in order to achieve the preservation of the throughput of a link to an application [2].

Conventional adapter architectures mainly accomodate networks that operate using variable size packets. Employment of ATM technology changes this network characteristic. The evolution of cell based ATM technology brings a new era in communications affecting the architecture and design of communication subsystems, which have

Asynchronous Transfer Mode Networks, Edited by Y. Viniotis
and R.O. Onvural, Plenum Press, New York, 1993

to provide connectivity to ATM networks. Although ATM technology is not spread today, it will clearly play a fundamental role in systems communications in the near future. So, it is necessary to provide ATM connectivity in the new emerging environment by designing adapters for communication subsystems that efficiently accomodate cell based, ATM, traffic in addition to traditional variable size packet traffic. ATM places some new architectural requirements on adapters due to its characteristic of interleaved packet transmission with fixed size cells. For example, packet interleaving coupled with the high speed of transmission links necessitates inclusion of fast packet segmentation and reassembly (SAR) circuitry. The high speed transmission links place strict requirements on the adapter architecture in order to achieve preservation of the link's throughput to the application. These requirements are placed on both packet data movement and protocol processing.

The strictest requirements in adapter architecture are placed by traffic of short packets, since all system operations have to be completed in time for every one of the short packets. For example, in a 100 Mbps network a 64-byte packet may arrive every 5.12 $\mu secs$; so, in an adapter accomodating 2 such networks at full speed, the reception (or transmission) operation of such a packet has to be completed within 2.56 $\mu secs$. The high cost of meeting such requirements in a subsystem accomodating a large number of network attachments frequently leads to a design that operates efficiently for long packets and can accomodate limited size bursts of short packets without data loss. In many environments this design decision is quite reasonable, since a large percentage of the traffic is composed of long packets. In ATM environments though it may be necessary to meet the strict requirements of successfully receiving every cell under heavy traffic in order to achieve good performance. In this paper, we analyze the effect of ATM and describe the necessary changes to a conventional adapter architecture in order to accomodate the high speed data movement required by multiple ATM attachments. The presentation is based on a prototype adapter that was designed and implemented to provide connectivity to high speed variable size packet networks such as FDDI [4]. The adapter achieves high performance for bridging and routing for relatively long packets.

The paper is structured as follows. Section 2 describes the architecture of the prototype adapter offering connectivity to high speed networks with variable size packets. Section 3 describes the Segmentation and Reassembly support required at the network attachment level and presents a cell admission policy to an attachment's elastic buffer that can lead to a high packet reception ratio. Section 4 introduces a partitioned memory management architecture that can be employed to increase system throughput. Conclusions appear in Section 5.

2 Basic Adapter Architecture

Figure 1 shows the general architecture of a prototype adapter for attachment either to an end system or to the backbone interconnection of a communication subsystem. The adapter provides connectivity to various networks through the specialized *Network Attachment* (NA) modules. The Network Attachments implement the protocols up to the MAC sublayer of the Data Link Control (DLC) layer and provide some elastic buffering and the interface to the remaining adapter modules. Packets incoming to (outgoing from) the adapter are stored in the *Data Memory* (DM) of the adapter. The *Processing Element* (PE) provides the higher layer protocol processing required on the packets received/transmitted by the system. All the adapter modules communicate over the *Adapter Bus*.

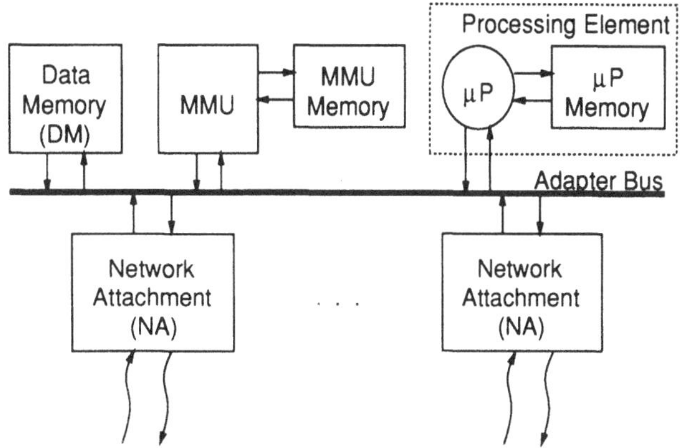

Figure 1: High-Speed Adapter Architecture

The packets stored in the adapter are typically organized in queues where each queue is associated with a specific incoming/outgoing data stream on a network link, packet priority, etc. The Data Memory is managed by a *Memory Management Unit* (MMU) in the following fashion: the memory is divided in a number of fixed size buffers. Buffers are linked together to form packets, and packets are linked to form queues. Managing the memory in this fashion allows for efficient memory utilization and low fragmentation in an environment where variable size packets from heterogeneous networks are accomodated. Since the Data Memory is shared among all attached networks, all the Network Attachment (NA) modules need the ability to manipulate the data structures in the memory accordingly, e.g. they have to indicate, directly or indirectly, the queue where a packet should be appended. This involves identification of the receiving/transmitting network, packet priority, etc. Memory management can be implemented in various ways, but in the prototype adapter we use a client/server model for memory management, where the MMU is the server and the Network Attachments are the clients. In this fashion, the MMU keeps track of the state of the Data Memory data structures and dispatches buffer addresses, receives queue id's, etc. from the Network Attachments upon their request. For example, when a new packet arrives at the adapter, the corresponding NA requests the address of a free buffer from the MMU; the MMU provides the address and then the NA starts moving packet data over the bus to the Data Memory in the designated buffer. As soon as the buffer is filled, the NA requests the address of a new buffer and moves data in it, while the MMU links the buffers together to form the packet. This process is iterated as long as necessary and finally the NA instructs the MMU to link the packet to the appropriate memory queue. All this communication between NAs and the MMU takes place over the bus.

The data structure information is also kept on the adapter in addition to the packet data and the higher layer protocol code. It has been demonstrated that a partitioned memory organization with 3 partitions: one for the packet data, one for the data structure information and one for the Processing Element's working space and program store, improves system performance relatively to the centralized memory approach; this occurs, because memory partitioning allows overlapping of various adapter operations such as data movement, data structure manipulation and protocol processing [2]. This partitioned memory organization is employed in the prototype adapter:

- the Data Memory is dedicated to storing only incoming/outgoing packet data;

- a specialized *Memory Management Unit* (MMU) with its local memory (MMU Memory) is responsible for the data structure manipulations;
- the Processing Element uses its private Local Memory (μP Memory) for the protocol code and working space.

Adapters employing the architecture depicted can be used to meet the requirements of available traditional LANs such as Ethernet and Token Ring as well as high-speed network links such as FDDI and T3 links. ATM networks need special consideration for attachment to the adapter. The reason is that ATM networks have some fundamental differences with the traditional Local Area and Wide Area Networks, due to Segmentation and Reassembly (SAR) of packets in fixed size cells, and packet interleaving.

3 Network Attachment Organization

In variable size packet networks, there is only one packet being received at every time instant. So, a receiving Network Attachment needs to know only a single Data Memory address at every instant during a packet reception; this address corresponds to the memory location, where the next incoming byte of the packet should be stored in Data Memory. After a packet is fully received, the Network Attachment obtains a new address to store the next incoming packet. Cell based, ATM, attachments in contrast may be receiving multiple packets simultaneously, due to packet cell interleaving. In this case, the Network Attachment receiving a cell identifies the packet to which the cell belongs, obtains the corresponding Data Memory pointer for the identified packet and then stores the cell to the appropriate memory location (buffer). Since several packets may be arriving interleaved simultaneously, the Network Attachment needs to keep track of the addresses of all packets under reception. For this purpose, a Packet Address table is employed at the Network Attachment level, which stores the pointers to Data Memory for every data packet being received. The design of the memory storing this Packet Address table is crucial to system performance for 2 reasons:

1. the number of entries in the table bounds the number of successfully received interleaved packets;
2. the speed of the Packet Address table should match the cell arrival rate of the incoming transmission link.

For an attachment running at OC-3 speed, a new cell may arrive at approximately every 2.8 $\mu secs$ under heavy load. This allows use of large size conventionally available memory to implement the Packet Address table and meet the link requirements. Several thousands of entries are easily implemented at a low cost.

Excluding the Packet Address table, the adapter bus is one of the two main resources involved in packet (cell) data movement; the other main resource is the Memory Management Unit (MMU). The adapter bus is the centralized resource that accomodates the movement of packet data to/from all attached interconnects. As such, the bus needs to provide effective throughput, T_E, that is: $T_E = \sum_{\forall i} T_{E(i)}$, where $T_{E(i)}$ is the effective throughput of the i-th attached interconnect. In addition to the packet data movement, the bus is also used for the communication between Network Attachments and the MMU, and so the performance of the MMU, in terms of both response and processing time, directly affects the effective throughput achieved in packet data movement to/from the memory. This memory management overhead, the overhead of bus

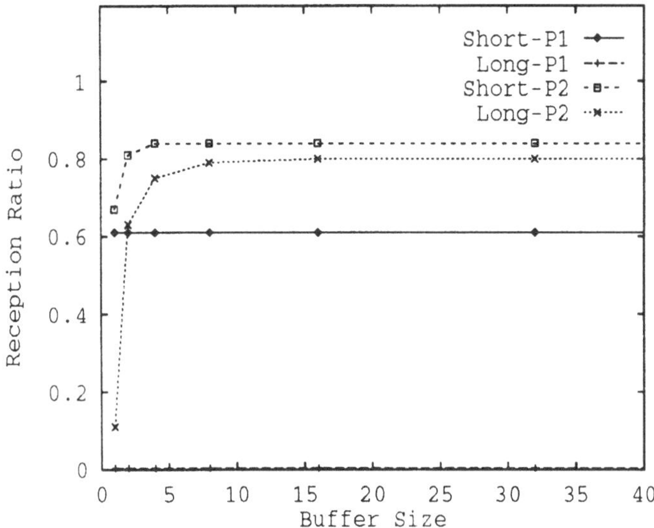

Figure 2: ATM Reception Ratio (ρ_S)

arbitration, and current technological limitations restrict the effective bus bandwidth. Furthermore, due to cost considerations, it is desirable to provide a large number of attachments on the bus. Overall, the above lead to designs and implementations where the available bus throughput does not match T_E, the cumulative bandwidth of the attached communication links. A critical issue in these designs is the effect of the limited bus throughput on overall system performance. In a network where each packet's bytes arrive continuously on the link, loss of consecutive bytes may result in loss of one or two packets, but the effect may be even more dramatic in an ATM environment: if k cells are lost, then k packets may be affected (or discarded). We analyze the effect of cell loss on system performance by simulation using as a metric the *Reception Ratio*, the complement of the packet loss probability, i.e. the fraction of successfully received packets: $\rho_S = \frac{P_R}{P_A}$, where P_R is the number of packets that are successfully received in the Data Memory of the adapter and P_A is the number of packets that arrive at the adapter on the attached link.

Figure 2 shows the Reception Ratio for the case of an ATM attachment where the bus throughput allocated to the attachment is 90% of the total link bandwidth. The results were obtained through simulation of an ATM Attachment receiving 1024 interleaved packets for variable elastic buffer sizes. To evaluate system behavior for different traffic pattern, the results for both short, $1 - 10$ cells, and relatively long, $40 - 100$ cells, packets are depicted. The results are shown for 2 different Network Attachment architectures.

The first architecture employs a straightforward cell admission policy, $P1$, to the attachment's elastic buffer where an incoming cell is accepted into the buffer as long as there is available space. The results, denoted "Short-$P1$", for short packets show that approximately 60% of the packets are successfully received, while almost all packets are lost (not successfully received) in the case of long packets as the plot denoted "Long-$P1$" indicates. Such an architectural approach will lead to a large number of cell/packet retransmissions placing a high overhead on the network.

The results are dramatically improved if a more sophisticated Network Attachment architecture is adopted, as the results denoted "Short-$P2$" and "Long-$P2$" demonstrate.

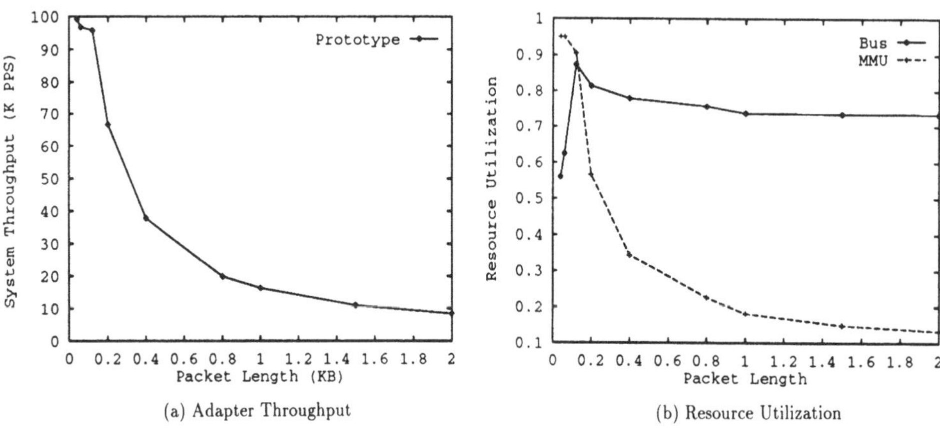

(a) Adapter Throughput

(b) Resource Utilization

Figure 3: Resource Utilization

The enhanced architecture employs the following admission policy $P2$ to the buffer: if a cell is discarded because no space is available in the elastic buffer, then all arriving cells belonging to the same packet are discarded, even when space is available in the buffer. Using this approach, at least 80% of both short and long packets are successfully received for sufficiently large buffer sizes as Figure 2 indicates. The Reception Ratio is very close to the maximal possible which is the 90% of the link, since the bus offers throughput equal to 90% of the link bandwidth. Minimal performance improvement is achieved though, when the offered bus throughput is not a high percentage (e.g., 90%) of the link throughput as simulation experiments have shown.

An important characteristic of the policy is its easy implementation. The Packet Address table is extended to include one extra bit per entry, the *Flush* bit, which has the following semantics: if it is set, then at least one cell of the corresponding packet has been dropped; the bit is cleared, otherwise. A simple finite state machine implements the policy in the following fashion: as soon as a cell arrives, a table lookup takes place in the Packet Address table to obtain the memory address where the cell will be stored into Data Memory (DM). The lookup returns the address and the Flush bit. If the Flush bit is cleared and there is enough space in the elastic buffer, the cell is stored, while in any other case, the cell is dropped and the Flush bit is set and written in the Packet Address table. This finite state machine is simple and adds minimal complexity to the Network Attachment design.

4 Distributed Memory Management

Although adoption of the previously described buffer admission policy achieves a high packet reception ratio in case the memory bandwidth per ATM attachment does not fully match the speed of the attached link, it is necessary to increase the Data Memory bandwidth in order to improve performance and achieve packet reception without flushing any data. The effective memory bandwidth is not only affected by the speed of the memory modules used, but by the performance of the memory management module as well. Consider, for example, reception of a packet; the packet data are moved to DM and then the packet is enqueued to the appropriate queue. A free buffer pointer is required by the Network Attachment to receive a newly arriving packet. Such a pointer

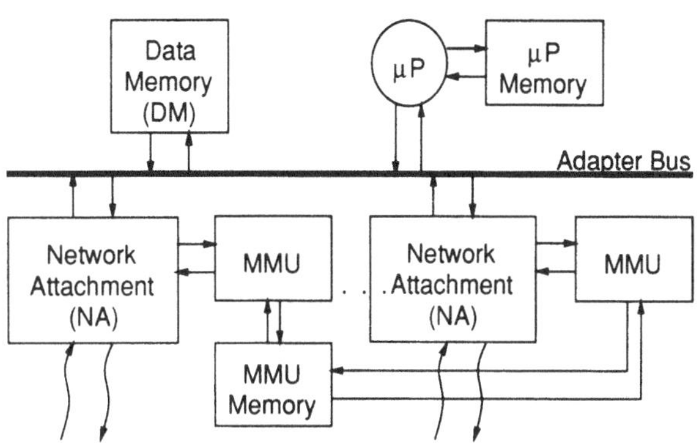

Figure 4: Distributed MMU Configuration

may not be available before the enqueueing process for the previously received packet is finished by the MMU. Even when caching and pointer prefetching is employed to overcome conflicts such as the above, this interaction cannot be avoided when short packets arrive in long bursts. This is clearly identified in Figure 3(a) where we plot the system throughput for the prototype adapter in packets per second as a function of packet length. The throughput is derived through system simulations with a high level model that closely approximates the behavior of the prototype system as has been verified. The results were obtained from an experiment where traffic composed of packets that arrived with zero interarrival time at one Network Attachment and were transmitted through a second one. No processing is considered on the packets. As the figure indicates, system throughput is limited to approximately $30 - 40$ *Mbps* for short packets, while for long packets it achieves almost 144 *Mbps*. The limitation is critical in accomodation of ATM traffic, which is composed of short cells only, and is due to the performance of the MMU as Figure 3(b) indicates showing the utilization of the bus and the MMU in the simulation experiments. The origin of this MMU performance limitation is that the MMU is shared among all Network Attachments and that all MMU communication takes place over the bus. We investigate two options for performance improvement in the system.

The first option is to detach the MMU from the bus, so that its communication does not interfere with the packet data movement. As Figure 5 demonstrates, this option provides performance improvement over the current design, but the number of short packets received by the system is low resulting in poor overall system performance. A more promising configuration is the adoption of a distributed MMU. Figure 4 shows the design of the ATM adapter with the distributed MMU approach for 2 attachments. The MMUs share the same memory storing the data structure information. This can be efficiently implemented for a moderate number of attachments operating in the range of $100 - 200$ Mbps, because there is minimal contention between the MMU requests; limited synchronization is required for enqueueing and dequeueing packets to the queues. Such an approach increases performance dramatically as Figure 5 demonstrates, where it is shown that throughput is approximately doubled for short packets in the simulated configuration. Figure 5 plots the throughput for all 3 system configurations, Regular adapter, Detached MMU and Distributed (Dual) MMU, in both measures, Packet per Second and Megabits per Second.

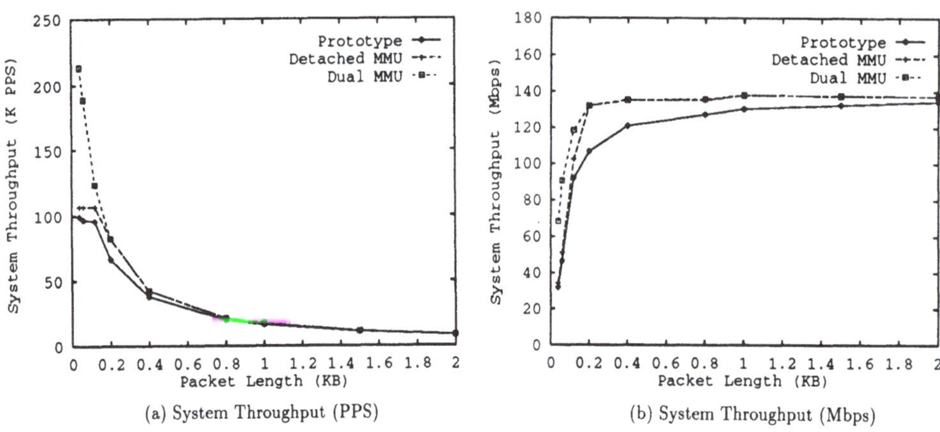

(a) System Throughput (PPS)

(b) System Throughput (Mbps)

Figure 5: Adapter Throughput

5 Conclusions

ATM technology places strict performance requirements on the architecture and design of network attachments due to its fixed size cell transmission of packets, which arrive interleaved at target adapters. Special considerations have to be taken to accomodate such traffic. We identified the required changes in the architecture of an implemented prototype adapter designed for variable size packets in order to accomodate ATM traffic. We described the support for packet segmentation and reassembly and we introduced a policy for cell admission in an attachment's elastic buffer whose positive effect on the number of successfully received packets has been demonstrated through simulations. Furthermore, a distributed memory management architecture was introduced that leads to communication subsystems that achieve high system throughput.

References

[1] IEEE, editor. *802.5: Token Ring Access Method.* IEEE, New York, 1985.

[2] H. E. Meleis and D. N. Serpanos. Designing Communication Subsystems for High-Speed Networks. *IEEE Network Magazine*, 6(4), July 1992.

[3] R. M. Metcalfe and D. R. Boggs. Ethernet: Distributed packet switching for local computer networks. *Communications of the Association for Computing Machinery*, 20(7):395–404, July 1976.

[4] F. E. Ross. FDDI-A Tutorial. *IEEE Communications Magazine*, 24(5):10–15, May 1986.

SMDS AND FRAME RELAY :

TWO DIFFERENT PATHS TOWARD ONE DESTINATION,

BROADBAND COMMUNICATIONS

Jean-Pierre Bernoux

Alcatel Network Systems, Inc.
2912 Wake Forest Rd.
Raleigh, NC 27609

LAN interconnection over long distances, the first Broadband service to become available on public network, can be done in two different ways: using SMDS or Frame Relay. These two technologies are emerging at the same time in the public network and operators are now ready to commit large amount of money as well as their reputation to back either (and sometimes both!) technology.This may explain why it is so difficult to get a fair comparative assessment of the two standards: it's too easy to insist on the absence of service standards in Frame Relay if you are on the SMDS side, or to scorn the complexity of SMDS if you are on the Frame relay side. But as we are witnessing the emergence of a huge data communication market, heading clearly toward Broadband communications, it is important that the choices made by all parties involved: users, operators and vendors, be based on knowledge rather than emotion.

DOES A COMPARISON MAKE SENSE ANYWAY?

Presently SMDS is based on the DQDB technology standardized as IEEE P802.6. It is planned that in the future it will rely on BISDN/ATM. The relationship between SMDS and 802.6 is easy to explain: as a public service, SMDS needs a standard interface between the Network and the Customer Premises. As the long term goal is to provide SMDS over the Broadband Network, the choice of 802.6 is a logical one. There is indeed a strong commonality between 802.6 and the ATM standards: both segment data into 48 Octets cells for transmission and switching over the network and both use a similar 5 Octets cell header for this purpose.

What should be compared to Frame Relay is the 802.6 standard. Both standards are defining ways to transport variable length protocol data units over public or private networks. ...But comparing Frame Relay and 802.6 is of little help because contrary to its relatives 802.3, .4 and .5 or to its cousin FDDI, 802.6 has no identified market nor have 802.6 prod-

Asynchronous Transfer Mode Networks, Edited by Y. Viniotis
and R.O. Onvural, Plenum Press, New York, 1993

ucts been announced. On the other hand, both Frame Relay and SMDS have vendors and network operators committed with product deployment in progress and the promise of seamless national and international service soon. As a result, a confrontation between the two contenders for LAN interconnection is going on now, on the maket place. As most market sensitive issues, this confrontation is handled by the media in a way that favors more salesmen behavior than impartiality. A comparison between Frame relay and SMDS that is based on an objective look at the standards indeed not only makes sense but is also needed.

TECHNICAL ASPECTS OF SMDS AND FRAME RELAY

THE SWITCHING TECHNIQUE

The first major difference to be noted is that, while Frame Relay relies on a network with Permanent Virtual Circuits (PVC), SMDS is connectionless. This divergence translates into two radically different switching and networking techniques that we need to describe more precisely.

If we have to interconnect two LANs, in a Frame Relay network, we must procede in two phases. First we need to setup connections in each node of the network that our traffic is going to traverse. Then an identification of the connection (called a DLCI for Data Link Connection Identifier) is put in front of each packet of data to be sent. This DLCI is a PVC identifier that is needed because the frames from different virtual circuits are statistically multiplexed on the same physical circuit. Statistical Multiplexing as opposed to Channel Multiplexing allows more flexibility in allocation of the bandwidth available on a link. As the data can no longer be identified by a channel on a multiplex, it is identified by a header placed in front of the frame. Figure 1 shows the structure of a frame. Figure 2 shows how DLCIs define virtual paths on physical links.

In SMDS/802.6, statistical multiplexing is the result of the connectionless mode of switching. Each datagram contains an address that uniquely identifies its destination. The format of the addresses in SMDS complies to the CCITT E.164 standard, i.e., it is structured like a telephone number. Each node analyzes this address and routes the corresponding packets accordingly. Figure 1 shows the format of an SMDS protocol data unit before segmentation.

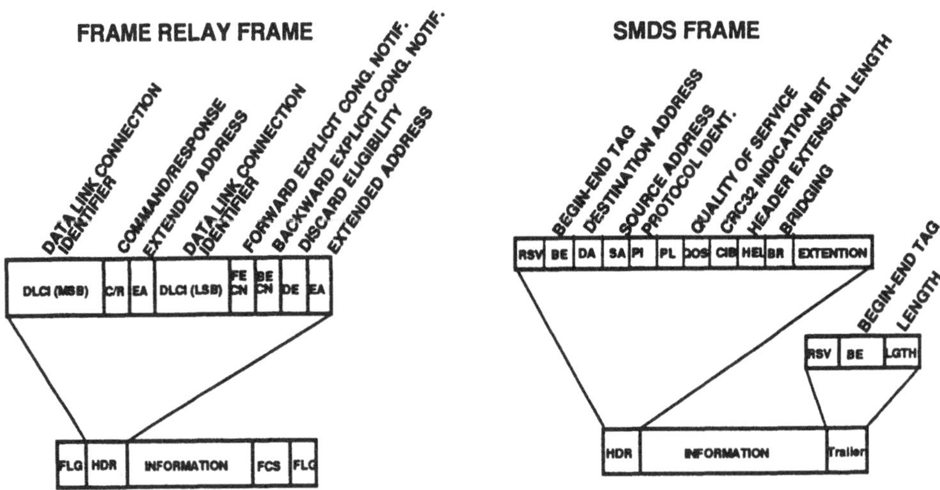

Figure 1. Frame Formats Comparison

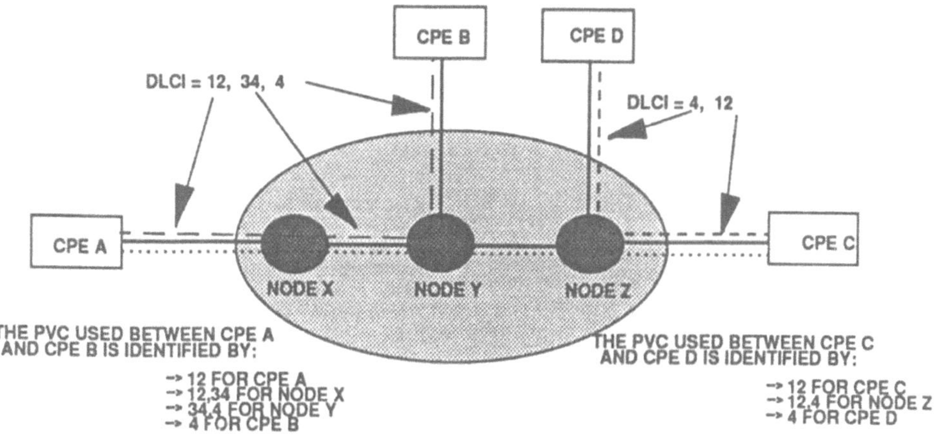

Figure 2. How DLCIs define Virtual Paths in Frame Relay

It may seem that the establishment of a connection in Frame Relay prior to sending data is a cumbersome constraint. But such connections are virtual connections, i.e., they consist only in memory that must be initialized in the nodes. In the absence of traffic, no other resource than a few octets of memory is used for each PVC in each node. It is thus possible to establish a large number of connections on a permanent basis which will provide enough connectivity for a user to reach all traffic destinations without constantly setting-up and breaking down connections.

A drawback of connection oriented networks is the complexity of the "call processing" involved in the call setup/tear down process. On the other hand, a PVC makes switching of packets in each node easier as Routing, i.e., the determination of the path to be followed by the packets, is performed once per PVC, at call setup.

While an E.164 address is more complex to analyze and route than a DLCI, the technology available today can perform these tasks at very high speeds. This technology relies on Application Specific Integrated Circuits (ASICs). As ASIC's performance/cost ratio improves steadily, the benefit of connectionless networks that do not carry the burden of signaling and call processing becomes more and more attractive.

A network with connections has two advantages over one that is connectionless : a guarantee of bandwidth during a call and a guarantee of message sequencing. However, when using PVCs, statistical multiplexing is allowed on the physical media shared by the virtual circuits and therefore there is a risk that data may be blocked in a node due to traffic congestion, just as in a connectionless network. As for the guarantee of message sequencing, it is taken care of by specific SMDS requirements.

If, in the future, high connectivity is needed, i.e., if a LAN has to communicate with hundreds of remote LANs, frame relay networks will have to introduce a way to establish connections dynamically. Presently PVC setup is an administrative process. No standard exists yet for Switched Virtual Circuits (SVC) but the CCITT standards for ISDN switched virtual circuits are likely to be selected. A mixed approach of PVC and SVC should provide both high connectivity and good performance, but such a solution has the following drawbacks:

• The choice of having a service switched rather than connected through PVCs is not trivial. The criteria to be considered like response time or load generated on the nodes require a level of coordination between users, node vendors and network operators that is unheard of in public networks.

• Depending on how extensively SVCs will be used, nodes deployed in a PVC context could prove to be short of the processing power required by SVCs.

THE TRANSMISSION OF DATA

A study of Data Communication techniques like X.25 tells us that a significant amount of processing performed besides switching consists of checking and restoring the integrity of the data transported on a hop by hop basis. Both SMDS and Frame Relay perform very limited checks and if anything goes wrong, the corrupted data is simply dropped. The reason for such ruthless behavior is that both SMDS and Frame Relay assume the use of very reliable media. As the probability of data loss or corruption is very low, why not rely on end-to-end mechanisms for recovery and save on processing at each intermediate node?

The physical lines used to support Frame Relay are T1 and sub-T1. The maximum speed on both access lines and trunk lines is 1.5 Mbps. SMDS relies on two standards: T1 for access lines and T3 for both access and trunk lines. The use of SONET STS-3c is also planned by 802.6.

Nothing however prevents Frame Relay from standardizing media at higher speeds. However, this has not been done yet for three reasons:

• Marketing studies show that currently 70 to 80% of LAN interconnections can be performed more or less effectively at T1 speed.

• Frame Relay is intended for deployment in the present network where the DS1 format is the main trunking standard (even when DS3 or SONET are used, in the present network these Broadband media are usually carrying multiplexes of T1s).

• Until now, the Frame Relay standards have focussed on the access rather than on the Node to Node Interface (NNI) where higher speeds are most needed to alleviate congestion problems.

Nothing prevents either SMDS from supporting sub-T1 speeds. An activity is going on in the SMDS Interest Group to standardize such an interface also called "Frame based interface" because it takes directly variable length SMDS Protocol Data Units. It is likely that this interface will use ... Frame Relay as a carrier!

SEGMENTATION AND REASSEMBLY

The typical unit of data handled in data communications is variable length, in the following we refer to it as a "message". With Frame Relay, the CPE presents messages to the network in one piece, regardless of their length. In SMDS/802.6, the CPE segments the messages into fixed length cells as described in Figure 3. Cells from different origins are multiplexed during transport and are switched individually within the nodes.

The practicality of segmentation is questionable in a network carrying asynchronous data only. The reasons segmentation was chosen in 802.6 are twofold:

• 802.6 will provide for the transport and switching of isochronous data, i.e., data that is sensitive to delays introduced by the network. Such data that is typically sampled at 8 Khz should not be delayed by large pieces of asynchronous data competing for transport. An efficient way to deal with this constraint is to multiplex cells of limited length (48 Octets for 802.6 and ATM) and to give a higher priority to the isochronous cells.

• To simplify the interworking between 802.6 and ATM.

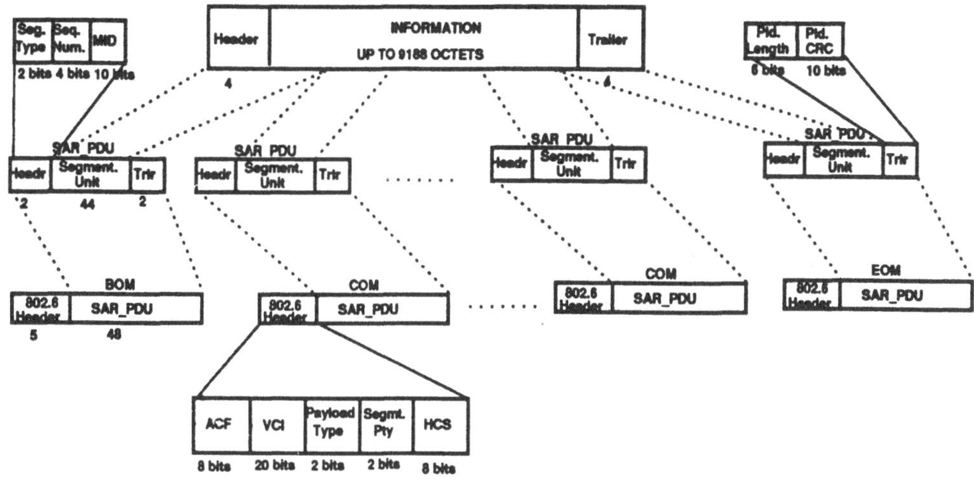

Figure 3. SMDS : Segmentation of SIP L3_PDU

Only the last point is relevant for SMDS as the transport of isochronous data is not supposed to be part of the SMDS service. Compared to Frame Relay, SMDS is penalized by the additional complexity of the segmentation and reassembly process. This penalty is justified by the objective of simplifying the interworking with future B-ISDN networks.

We said earlier that the data is sent in one chunk in Frame Relay networks. This might not be always true: the default maximum frame size is set at 262 octets in the standard; as this value is not adequate for LAN interconnection, each supplier has selected a larger maximum frame length for their implementation, these values range presently from 1600 to 8K octets. This leads to two observations:

• With a 1600 octets frame, it is necessary to introduce a Segmentation and Reassembly (SAR) layer for all packets larger than 1600 octets coming from a CPE.

• Between two network nodes using different maximum size of frames, a SAR layer is also necessary.

CUSTOMER PREMISE EQUIPMENT

The CPE interface is the strong point in both standards. There are good reasons for that: it is unlikely that end users would buy proprietary CPEs that would lock them in with a node vendor or network operator. On the contrary, a successful standard promotes lower prices through competition and preserves the user's investment.

The simplicity of the Frame Relay interface is an advantage. It is possible to turn a router with a T1 interface into a Frame Relay CPE by a software or firmware retrofit.

SMDS CPEs are also based on routers but as the SAR function of 802.6 was too complex to handle, an external SMDS DSU had to be introduced. As it appeared quickly that there would be more router vendors willing to provide SMDS service than DSU vendors, the standardization of the router-DSU interface became a necessity. This was achieved in 1992 by the SMDS interest group and the standard is known as the DXI for Data eXchange Interface. This interface has the particularity to be "Frame based" and to rely on the SMDS L3_PDU.

A further advantage of Frame Relay is a better use of bandwidth. The short Frame Relay header introduces little overhead compared to the overhead created by the larger header and

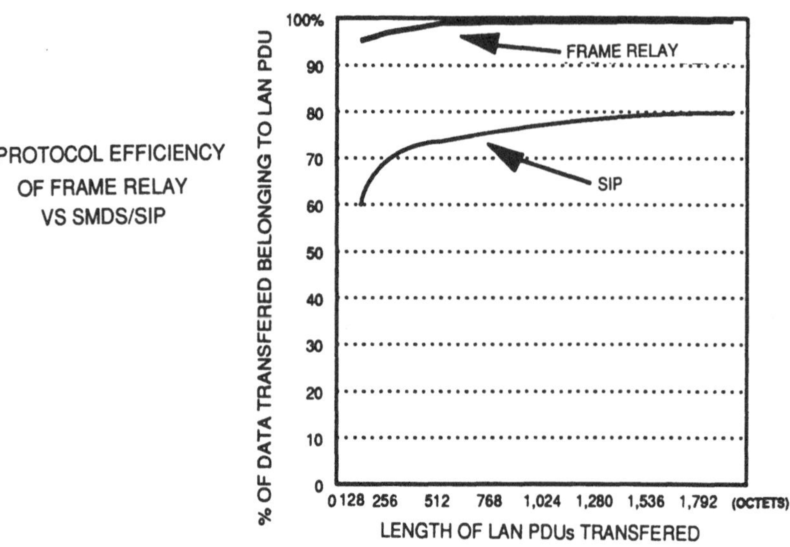

Figure 4. Protocol Efficiency Comparison

the cell structure of SMDS. Figure 4 gives a comparison between the protocol overhead of Frame Relay and SIP, the equivalent SMDS Interface Protocol. This advantage is challenged by the SMDS frame interface used for low speed access as explained earlier.

SERVICES BEYOND BASIC DATA SWITCHING AND TRANSPORT

SMDS End user features

***Address validation*:** The network ensures that the sender cannot give a fraudulent source address.

***Group Addressing*:** a group address used as a destination allows the multicasting of the information to all members of the group. This feature is similar to the "Broadcast" capability of LANs. It is used extensively by some protocols for routing management and address resolution information.

***Address screening*:** Performed at destination on Source Addresses, allows delivery from a predefined list of senders only. Performed at source on Destination Addresses, allows transmission to a predefined list of destinations only. In summary, Address Screening allows the implementation of Logical Private Networks on the public network.

***Access Classes*:** provide customers with different traffic characteristics. Although the capacity of the physical medium can be used entirely by the CPE for bursts, the access class defines a limit on the average rate of data transfer allowed over a longer period.

Five Access Classes have been defined for DS3, ranging from 4 to 34 Mbps. For each Access Class, the maximum DS3 speed can be used for the first 9188 octets transfered in a burst. Beyond that, the network Operator can drop the data in excess of the subscribed rate.

Access Classes are useful for service provisioning. The operator can dimension his network based on the subscribed bandwidth rather than on the througput of the access media. Access Classes also provide flexibility in charging. More precisely, a user has the possibility to subscribe based on the power of his CPE (i.e., the maximum bandwidth his CPE can use) rather than on the bandwidth of the access medium. In the regulated environment, Access Classes are mandatory for the RBOCs to be allowed to charge less than the full DS3 access rate to subscribers who need only a fraction of the 45 Mbps bandwidth.

Frame Relay end user features

While CCITT and ANSI standards are not focusing on end users, the following optional features have been defined in these standards:

• ***Explicit congestion notification*:** Nodes can send forward and backward congestion notifications for given PVCs.

• ***Implicit congestion notification*:** This is a mechanism that can be implemented in the CPEs, to throttle the traffic upon certain criteria (delay or repetition) implying congestion in the network.

In 1990, four vendors (cisco, DEC, NTI and StrataCom) agreed on a proposal for extensions to the ANSI/CCITT standard. This proposal defines the Local Management Interface (LMI) as consisting of two types of extensions:

• **Common extensions**, on which all sponsors agree including:

- A *common format* for the management messages exchanged between the CPE and the Access Node

- A *keep alive sequence*, intended at checking that both extremities on an access line are operational

- A *PVC Status*, allowing to notify the user device of the configuration and status of an existing PVC.

• **Optional extensions**, on which several sponsors agree, that mainly consist of:

- A *global addressing mode* allowing under some conditions to identify the source of a frame (Note that in the Frame Relay header, the identification of the sender is not available).

- A *multicast capability* that allows sending a frame to all members of a multicast group.

- A *simplified flow control* scheme based on the XON/XOFF mechanism.

Since the introduction of this proposal, more than thirty vendors have regrouped behind the initial four in support of the LMI. This group is now known as the "Frame Relay Implementors' Forum".

SMDS Network Operator Features:

***Operations, Administration and Maintenance*:** Many features are defined for OA&M. The objective is to standardize the interface with service providers' Operations Systems (O.S.), which will provide a unified view, independent of vendors and switch equipment. The main areas covered are: Memory Management, Maintenance, Traffic Management, Network Data Collection, Customer Network Management.

Billing: Provides a description of usage measurements for billing, performance and operations related to billing.

Performance objectives are defined in several areas: Maximum transit delay for a packet; probability of loss, error, misdelivery; service availability... are a few examples.

Frame Relay Network Operator Features

There is nothing equivalent to the SMDS OA&M and Billing features for Frame Relay. Some of these capabilities are needed to operate a network and are available in existing products. Presently each Node Vendor is defining its own proprietary solutions in the absence of standards. This approach is fine as long as the nodes of a network are from the same vendor. As it's difficult to imagine a country like the USA covered by a network built by a single vendor, there will indeed be interworking problems between the different Frame Relay networks deployed. In a public service context, these problems could grow in a combinatorial manner.

ABOUT CONGESTION

Since nothing in life is free, especially not in Telecommunication, there is a price to pay for the dynamic bandwidth capability offered by both SMDS and Frame Relay. The price is the necessity to drop information in the network in case of congestion. Without control, congestion can create an avalanche effect leading to a collapse of network's throughput. The culprit is the transport layer of the protocols in the CPEs which keep asking for retransmission of data until they get through.

Frame Relay and SMDS have mechanisms described in the standards which deal with congestion notification. The actions to be taken by each node in case of congestion are not specified. There is no guarantee of graceful reaction of either type of network in case of congestion.

The most efficient way to deal with congestion is to throttle traffic at the access point. This is done differently in each standard:

• Frame Relay relies on the cooperation of the CPE which receives an explicit congestion notification. The access node can also mark frames with a "discard eligibility" mark (see Figure 1) that indicates to the network which frames can be discarded in priority in case of congestion. The frames so marked are usually those that correspond to traffic beyond the "Committed Information Rate" (CIR). The CIR is a subscription parameter that the user selects based on his estimated traffic needs.

• SMDS relies on the Access Class which corresponds to the rate of traffic and burstiness subscribed to by the user. In case of congestion the user is not notified explicitly, nor is he expected to reduce his traffic. The traffic measurements specified in SMDS allow the acquisition of information on traffic patterns. This information in turn should help improve dimensioning of the network by service providers.

CONCLUSION

Although both SMDS and Frame Relay initially aim at the same market of LAN interconnection over wide areas, these standards are the result of radically different approaches to this market:

• Frame Relay is mainly backed by the vendor community. It focuses on providing a basic service quickly, in continuity with existing products and networks, for a minimum cost.

• SMDS is mainly backed by the seven Regional Bell Holding Companies (RHC's) through their common research concern, Bellcore. SMDS focuses on early introduction of Broadband services in the public network.

From a standard perspective, the strength of SMDS is that it is defined by a coherent organization, Bellcore. A potential weakness of SMDS is that the logic followed by Bellcore is intended at satisfying first its clients, the RBOCs. Will the requirements expressed by the RBOCs satisfy the needs of the end user? A first answer to this question can be found in the activities of the SMDS Interest Group which is quickly filling the gaps left open by Bellcore. Two examples are the DXI interface which allows to standardize the interface router-DSU and the introduction of SMDS service at sub-T1 speed.

The strength of Frame Relay is its simplicity and its adaptation to the existing networks and equipment. Its weakness lies in the area of public networks where nothing actually is simple, particularly when it comes to manage the network. In theory, the Frame Relay standards could be expanded to cover SMDS services and to allow higher speeds. Such an outcome may seem unlikely within a reasonable timeframe given the concensus approach required by ANSI and CCITT committees. The situation is not hopeless however, the example of the LMI initiative shows that, when a significant number of key players are able to get organized behind a concrete proposal, the committees have little choice but to change their pace... and follow.

The main threat for Frame Relay is a combination of slow commercial take-off and acceleration of the ATM momentum which might close the "window of opportunity" that was quite open a year ago. This threat is further exacerbated by the move of SMDS toward low speed, an area where the size of the market is substantial and where Frame Relay was unopposed initially.

The main threat for SMDS is that, to be successful, it has to be available nationally which cannot happen without the cooperation of the Interexchange Carriers (ICs). Unfortunately for the RBOCs the ICs are also competing to provide data services directly to end users. The ensuing arm twisting between RBOCs and ICs might result in delays for providing InterLata service for SMDS.

REFERENCES

Frame relay standards

ANSI T1.606, CCITT I2xy: Service description

ANSI T1.606 addendum, CCITT I.3xz: Congestion Management Strategy

ANSI T1.618, CCITT Q.922: Core Aspects of Frame Relay Protocol.

ANSI T1.617, CCITT Q.93x: Signaling specification for Frame Relay Bearer Service.

CCITT Recommendation I.122: Framework for Providing Additional Packet Mode Bearer Services, Blue Book

SMDS Standards

TR-TSY-000772: Generic System Requirement in Support of SMDS Service

TR-TSY-000773: Local Access System Generic Requirements, Objective and Interfaces in Support of SMDS Service

TR-TSY-000774: SMDS Operation Technology Network Element Generic Requirement

TR-TSY-000775: Usage Measurement Generic Requirements in Support of Billing for SMDS Srvice

TA-TSV-001059: Inter-Switching Interface Generic Requirements for SMDS Service

TR-TSV-001060: Exchange Access Generic Requirement for SMDS Service

TA-TSV-001061: Generic Requirement for Operations Technology for Inter-Switching System Interface (ISSI) and Exchange Access Interface (XA_SMDS) and Usage Measurements for Billing in Support of XA-SMDS

TA-TSV-001062: Generic Requirements in Support of Customer Network Management for SMDS Service.

TA-TSV-001063: Operations Technology Network Element Generic Requirements in Support of XA_SMDS

SR-TSV-002198: Support of Inter-Carrier Aspect of SMDS in a Multi-Switch Network

ATM SYSTEMS IN SUPPORT OF B-ISDN, FRAME RELAY, AND SMDS SERVICES

Paul Holzworth

Fujitsu Network Switching of America, Inc.
Raleigh, NC 27609

INTRODUCTION

New communication services such as Frame Relay, Switched Multimegabit Data Service (SMDS) and Broadband Integrated Services Digital Networks (B-ISDN) will require changes in the public data network. In the future, the public network will also need to respond to users' requirements for bandwidth-on-demand. Asynchronous Transfer Mode (ATM) technology can provide these capabilities and can provide a versatile network that can support many different services. New services such as Frame Relay and SMDS can be transported and switched by a public ATM network. This paper will address some of the issues involved in supporting these services on an ATM platform and will provide solutions that have been used in an existing ATM switch.

WHY ATM?

The design of the public communications network has been based on the requirement to carry voice traffic between analog telephones. With the arrival of digital T-carrier transmission systems, the public network began to evolve into a digital network. The digital network was intended to carry channelized voice data using 64 Kbps data channels. Digital switching systems were added to the network and voice could be digitally carried within the network without requiring additional analog/digital conversion. ISDN extended the digital network to the subscriber and provided the ability to provide both voice and data service to the subscriber. ISDN service can be expanded in 64 Kbps increments.

As workstations and computers become faster, high speed communications becomes more important. 10 Mbps Ethernet speeds are giving way to 100 Mbps FDDI speeds and supercomputers are already using 800 Mbps HIPPI. There are increasing demands on the public network to support data communication at higher speeds. In the past few years, Frame Relay and SMDS services have been introduced to accommodate data rates up to 45 Mbps. B-ISDN systems will support data rates to 622 Mbps and beyond.

ATM has been chosen as the technology for use in B-ISDN. It can support many types of services at a variety of speeds. This makes ATM particularly appropriate for use with multimedia services that require multiple data channels operating at different speeds. Frame Relay and SMDS services can also be provided by an ATM network. A unified approach to system maintenance, monitoring, and billing can be provided by a single system since ATM can support multiple services on a single system that may require multiple systems using a non-ATM approach.

Asynchronous Transfer Mode Networks, Edited by Y. Viniotis
and R.O. Onvural, Plenum Press, New York, 1993

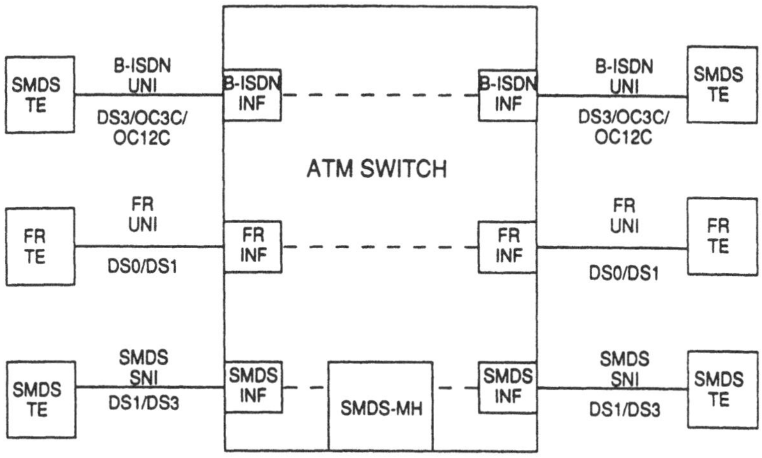

Figure 1. ATM switching systems can support a variety of interfaces.

B-ISDN

B-ISDN offers a set of high speed services based on ATM and SONET technology. Interface rates ranging from 45 Mbps to 622 Mbps have been standardized by ANSI and CCITT.[1,2] Bellcore has also provided a detailed series of Technical Advisories that list a set of requirements for B-ISDN.[3,4] These services provide a framework that can be used to support a variety of end-user applications. These applications include videoconferencing, transmission of HDTV, high speed digital data transmission and others. Some of the early trials of B-ISDN have consisted of the transmission of X-rays to remote locations and transmission of high speed data between the HIPPI interfaces of supercomputers.[5]

It is necessary to provide additional functionality above the ATM layer to provide usable interfaces to subscribers. This functionality is provided by the ATM Adaptation Layers (AAL). Separate AALs have been defined to provide connectionless and connection-oriented packet services and circuit emulation services[6]. Figure 2 shows the protocol stack that is used by an ATM switch to support B-ISDN. The terminal has a direct ATM-based B-ISDN interface.

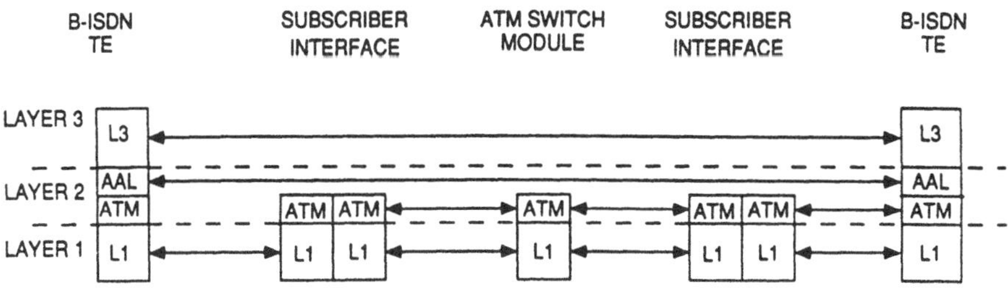

Figure 2. B-ISDN Protocol Stack

FRAME RELAY

Frame Relay is a connection-oriented packet data service. It is similar to X.25 packet services. Frame Relay is designed to operate at higher speeds and with lower protocol overhead than X.25 packet services. It is currently defined for interfaces operating at rates of 56 Kbps and 1.544 Mbps using DS0 and DS1 respectively. Future implementations of Frame Relay are also expected to operate at 45 Mbps over DS3. Initial implementations of Frame Relay operate over permanent virtual circuits but future implementations will be provided over switched virtual circuits.

Frame Relay is a data transport service that operates at layer 2 of the OSI reference model. Frame Relay uses variable length data packets. Each packet contains a field which is used to identify the virtual channel associated with the packet[7]. The virtual channel is identified by the Data Link Channel Identifier field (DLCI) in the packet header. The channel identified by the DLCI is unique on a Frame Relay link, but is local and only has significance within the link. A Frame Relay connection that spans multiple links may have a different DLCI for each link that together define the single connection. This is analogous to the virtual channel identifier used by ATM to uniquely identify each connection on an ATM link. This makes it simple to map Frame Relay connections onto ATM by mapping each DLCI onto an ATM virtual channel.

A Frame Relay interface can support multiple connections as long as the sum of the data rates does not exceed the interface rate. Networks are usually engineered with a safety margin so the sum of the data rates is only a fraction of the interface rate. Frame Relay supports two parameters for traffic policing. The Committed Information Rate(CIR) is the data transmission rate that the network will guarantee to provide for the connection. (This "guarantee" is actually defined between the user and the network provider as a mutually agreed probability of delivery and is not an absolute guarantee.) The Burst rate excess (Be) parameter is the maximum data rate that the network will accept for a connection. Any data that is submitted in excess of Be is immediately discarded. Data submitted in excess of the CIR but below the Be, will be marked by the network as eligible for discarding by setting the Discard Eligible indicator in the frame. This is an indication to the network that this data should be dropped first if the network experiences congestion or capacity overload. The Discard Eligible indicator may also be set by the user to indicate that the marked frame should be dropped by the network in preference to a frame that is not marked. Similar concepts exist in ATM. ATM cells are submitted to the network at a subscribed-to data rate and excess cells may be marked by the network by setting the Cell Loss Priority bit. The user may also set the CLP bit to indicate that some ATM cells should be dropped in preference to other ATM cells. The Frame Relay Discard Eligible indication can be mapped to the ATM CLP indication when transporting Frame Relay through an ATM network.

Frame Relay to ATM conversion is performed at the subscriber interface card. Figure 3 shows the protocol stack for a Frame Relay connection across an ATM network. The subscriber interface card terminates the physical layer (DS0, DS1, or DS3) and provides the performance monitoring and testing capabilities required of the physical interface. Figure 3 shows the mapping of data from Frame Relay into ATM. The data link core can be supported by using either AAL 3/4 or AAL5. Initial implementations have used AAL 3/4. The subscriber interface terminates the Frame Relay UNI protocol. It is therefore responsible for performing all of the necessary data collection functions such as traffic management, performance monitoring, and billing data collection. This information is then sent to the operations and maintenance system of the ATM switch.

SMDS

Switched Multimegabit Data Service (SMDS) is a connectionless service that is used to provide wide area connectivity for data terminals and LANs. Interfacing connectionless

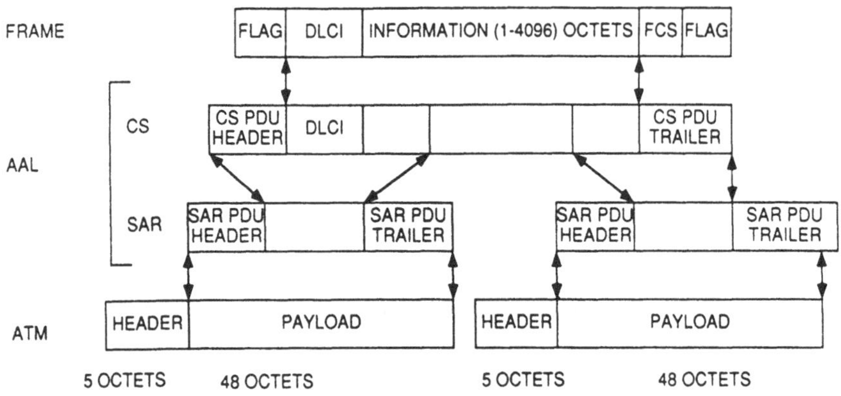

Figure 3. Frame Relay to ATM mapping

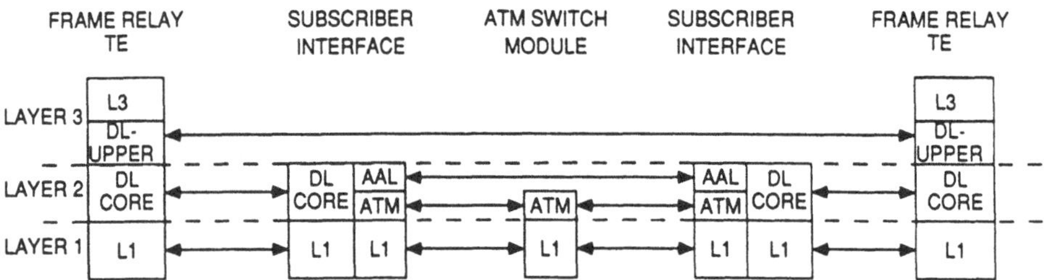

Figure 4. Frame Relay protocol stack

SMDS with a connection oriented ATM network presents unique challenges. ATM switching systems require a call to be set up before data may be transmitted. Connectionless SMDS packets are sent into the network without regard for call setup. In order to provide the ability to send connectionless data over a connection oriented network, we have implemented a device called a message handler. The message handler acts as an SMDS data server and performs address mapping from the SMDS destination address to a subscriber interface in the network.

Figure 5 shows how SMDS data is mapped into ATM cells[8]. SMDS Level 3 Payload Data Units (L3_PDUs) may contain up to 9188 octets of user data. Each L3_PDU is segmented into 44 octet segments that are mapped into SMDS Level 2 Payload Data Units (L2_PDU). The L2_PDU contains a header field that indicates that the L2_PDU is the Beginning of Message(BOM), Continuation of Message(COM), or End of Message(EOM) of the segmented L3_PDU. The L2_PDUs are then mapped into the 48 byte payload of an ATM cell.

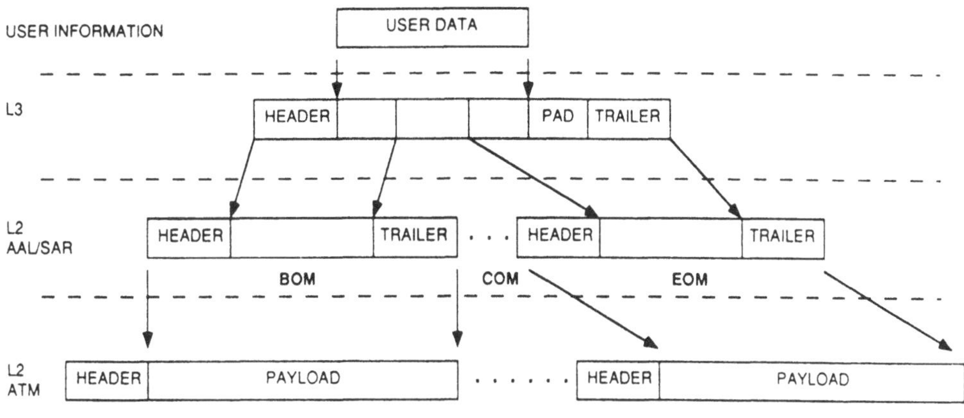

Figure 5. SMDS data mapping.

The L2_PDU contains a 10 bit message identification field used to identify a particular series of messages made up of different segments (BOM,COM, and EOM segments). The ATM cell also contains a 24 bit Virtual Path Identifier / Virtual Channel Identifier (VPI/VCI) field that is used to uniquely identify each connection on each physical link. SMDS addressing information is contained in the L3_PDU in the form of a source address and a destination address. Within an ATM switch, all routing is performed at the ATM cell level using the VPI/VCI field. Higher layers in the protocol stack are not processed. This may be seen in Figure 6. Once a call has been set up, cell routing occurs automatically. In order to provide SMDS service over ATM, each subscriber interface is connected to a message handler with a Permanent Virtual Circuit (PVC). Multiple message handlers may be used to support more subscribers. The ATM switching system will take all cells from the subscriber interface and route them to the corresponding message handler. The message handler extracts the destination address from the SMDS L3_PDU. The message handler examines the destination address and looks for it in a local address table. If the address is found, the message handler forwards the L3_PDU as ATM cells to the message handler that is connected to the destination subscriber interface. The destination message handler converts the L3_PDU into SMDS L2_PDUs and transmits them over the SMDS SNI to the subscriber. If the address is not found, the message handler forwards the L3_PDU to a special gateway message handler. The gateway message handler functions as a gateway to other switching systems and is connected to other gateway message handlers in other switching systems.

Careful consideration must be given to message processing. The L3_PDU is pipelined within the SMDS switching system to avoid the large amounts of delay that could be introduced into the network if the message handler waited for the entire L3_PDU to be assembled before beginning processing. (1028 bytes at 1.544 Mbps DS1 rates equates to 5.3 ms reassembly delay). It is not necessary for the message handler to reassemble the complete L3_PDU before starting to process it. The first L2_PDU (BOM) contains the header for the L3_PDU. It is therefore possible to determine the source address and the destination address of the L3_PDU from the first L2_PDU. After the BOM, the message handler can forward each COM L2_PDU to the destination message handler. The message handler detects the end of the L3_PDU when the EOM L2_PDU is processed. If an error has occurred in the transmission of the L3_PDU, the final L2_PDU is not forwarded, forcing an error condition at the receiving terminal equipment.

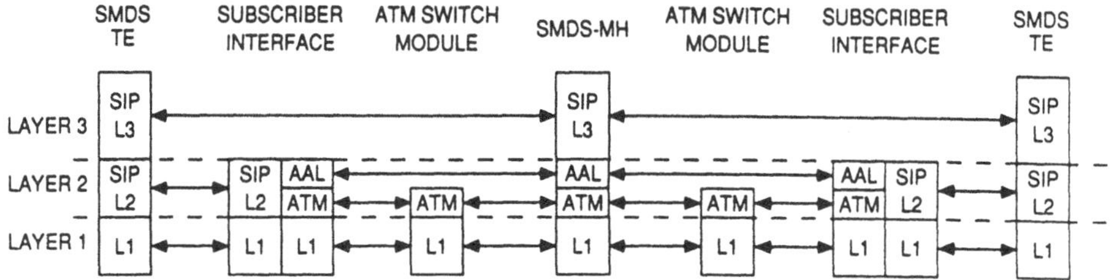

Figure 6. SMDS Protocol Stack.

CONCLUSION

Switching systems that use ATM switching technology can support a variety of services on a single platform. This paper has presented one approach for supporting Frame Relay, SMDS and B-ISDN from a single switching platform. Commercial ATM switching systems based on this approach are now available.

REFERENCES

1: ANSI T1S1, T1S1 LB91-05, Draft Broadband ISDN User-Network Interfaces: Rates and Formats Specifications

2:CCITT Draft Recommendation I.432, "B-ISDN User-Network Interface - Physical Layer Specification", Study Group XVIII, Geneva, June 1992

3:Bellcore, TA-NWT-01112, Issue 1, August 1992, "Broadband ISDN User to Network Interface and Network Node Interface Physical Layer Generic Requirement"

4:Bellcore, TA-NWT-01113, Issue 1, August 1992, "Asynchronous Transfer Mode(ATM) and ATM Adaptation Layer (AAL) Protocols Generic Requirements"

5: K. Chipman et al, "Medical Applications in a B-ISDN Field Trial, IEEE Journal on Selected Areas in Communications, September 1992,

6: CCITT Recommendation I.362, B-ISDN ATM Adaptation Layer Functional Description, Geneva 1991.

7: ANSI T1.618 Integrated Services Digital Network (ISDN) - Core Aspects of Frame Protocol for use with Frame Relay Bearer Service

8: Bellcore, TR-TSV-000772, "Generic System Requirements in Support of Switched Multi-Megabit Data Service", Issue 7 1, Bellcore Technical Reference, December 1991

Approaching B-ISDN: An Overview of ATM and DQDB

C. Bisdikian, B. Patel, F. Schaffa, and M. Willebeek-LeMair

IBM T.J. Watson Research Center
P.O. Box 704
Yorktown Heights, N.Y. 10598

Abstract

Asynchronous Transfer Mode (ATM) is the internationally agreed upon technique for transmission, multiplexing, and switching in a broadband network. It is designed to support the integration of high quality voice, video, and high speed data traffic. Although much work is underway in the definition, design, and development of ATM, many important components of the ATM network have not been solved. ATM is a transfer mode suitable for implementing broadband networks (WAN, MAN, and LAN) but not a network architecture in and of itself. Although, the origin of DQDB MAN is unrelated to ATM, the work by IEEE 802.6 that lead to the standardization of the DQDB will be the introductory step to ATM and B-ISDN. The relationship between ATM and DQDB is explained and a detailed discussion on the open and challenging problems that remain to be solved for ATM is presented.

1 Introduction

Asynchronous Transfer Mode (ATM) [1, 2, 3] is the internationally agreed upon technique for transmission, multiplexing, and switching in a broadband network. ATM is a CCITT (*Comité Consultatif International Télégraphique et Téléphonique* - the International Consultative Committee for Telecommunications and Telegraphy) standard for broadband ISDN (B-ISDN). It is designed to support the integration of high quality voice, video, and high speed data traffic. To the end-user, it provides the ability to transport connection oriented and connectionless traffic at constant or variable bit rates. It allows for allocation of bandwidth on demand and intends to provide negotiated *Quality-of-Service* (QoS). To a network provider, it enables the transport of different traffic types through the same network.

Much work is underway in the definition, design, and development of ATM, however, many important components of the ATM network have not been solved. Multiple classes of service are defined for ATM with constant or variable bit rates, connection oriented and connectionless modes of transport, and possible timing requirements between source and destination. Supporting all of these parameters places severe requirements on the network in order to guarantee satisfactory performance and QoS.

Asynchronous Transfer Mode Networks, Edited by Y. Viniotis
and R.O. Onvural, Plenum Press, New York, 1993

ATM is a transfer mode suitable for implementing broadband networks (WAN, MAN, and LAN), but not a network architecture in and of itself. Therefore, an ATM based network may be implemented using many different topologies (switch based or shared media), different media access control mechanisms, physical layers, etc. The definition of ATM Local Area Networks (LANs) is actively being pursued by the ATM forum—a consortium of computer and network manufacturers. In the evolutionary process towards ATM (B-ISDN), DQDB networks will have a significant role.

In contrast to ATM, DQDB is a complete Metropolitan Area Network (MAN) architecture. Although the origin of DQDB MAN is unrelated to ATM, the work by IEEE 802.6 that lead to the standardization of the DQDB might be considered an introductory step to ATM and B-ISDN[4]. The Switched Multimegabit Data Service (SMDS)[5, 6] provided by the RBOCs will be DQDB-based. From the outset of the IEEE 802 Project, more than a decade ago, it was well recognized that a standard for a communications network that would cover metropolitan size areas had to be developed. The IEEE 802.6 standards committee had agreed that such a MAN would ultimately provide integrated services (e.g., data, voice, and video) on a common high-speed shared medium network.

The 802.6 MAN borrowed ideas from the other 802.X LAN [7] projects (e.g., multi-access of a shared communications resource), however, the DQDB MAN is not a LAN. Several key differences exist. First, due to its geographical coverage ($\simeq$ 50Km in radius), the *media access* (MAC) *protocol* of a MAN should be relatively insensitive to the physical size of the network. Hence, simple extensions of the other 802.X MAC protocols are not acceptable solutions due to their distance limitations. Second, one of the first applications envisaged for a MAN is the interconnection of geographically dispersed LANs. This makes it desirable that a MAN should operate at speeds well above that of the other 802.X LANs. Third, the stringent delay requirements of real time traffic (e.g., voice, video) cannot be easily satisfied by the other 802.X LANs. Finally, once again due to its size, a MAN is expected to be a *utility* that is centrally administered and operated unlike the other LANs that are privately owned and operated. Hence, billing, maintenance, and security are important issues in a MAN.

In this paper we begin with a brief review of the key aspects and characteristics of ATM and DQDB. Specifically, in Section 2, we discuss the philosophy behind ATM and describe its layered organization In Section 3 we describe the origin of DQDB and the operation of its MAC protocol. The relationship between DQDB and ATM is discussed in Section 4. Section 5 includes a detailed discussion on the open and challenging problems that remain to be solved for ATM.

2 ATM Philosophy

The dramatic increase in network link speeds and the relative moderate increase in CPU processing speeds has required an overhaul in communication architectures. Whereas in the past the ratio (α) between the communication processing rate (per byte) and communication transmission rate were comparable ($\alpha \sim 1$), current transmission rates have led to a significant imbalance ($\alpha >> 1$). The ratio imbalance can only be overturned by either increasing the processing speed or decreasing the amount of processing required. Both approaches have been pursued. In the first case, special communication hardware has been introduced or dedicated communication processors have been incorporated into the system design. In the latter case, communication protocols have been streamlined to reduce overhead or to allow separate control and data paths through the system for fast data handling. Furthermore, the decrease in link bit error rates has diminished the need for "heavy duty" error correction and recovery mechanisms typical of traditional transfer and network layer protocols. These factors have all been crucial to the ATM philosophy.

The ATM philosophy is based on several basic principles. First, a multiplicity of services with varying characteristics and QoS requirements should be supported by the same underlying network. Hence, the network control should be simple to support low-delay requirements (by reducing processing overhead). Second, virtual channels are used to define and characterize communication paths through the network. This allows each communication path to be customized based on the desired service requirements and allows multiple virtual channels to share physical communication links to efficiently utilize network resources. Finally, the underlying mechanism which will enable the first two principles to be implemented across high speed links is the small fixed-size ATM cell. The cell facilitates high speed data processing in hardware in the end-stations and low-delay switching in the network nodes.

It is important that a broadband network support the continuous evolution of the underlying technology and support current and future services. Improved compression algorithms and advances in VLSI technology may reduce the bandwidth requirements of existing services. New services may emerge with unknown service requirements. ATM is designed to be flexible in adapting to a variety of service requirements. Furthermore, it efficiently utilizes network resources by sharing available resources through statistical multiplexing.

ATM is based on fast packet switching technology. The fast packet switching is accomplished by providing minimum functionality in the network, connection oriented transport, and small fixed size packets (cells) with reduced header functionality. Minimum functionality in the ATM network is achieved by moving error handling or flow control mechanism to the periphery of the network. A cell in error is dropped by the network upon error detection. Any need for reliable transmission is handled by the end-station. Such a solution is acceptable because low bit error rates are achievable with today's communication links. In case of packet-loss due to congestion, no special action (e.g., flow control) is taken by the network. Protection against packet-loss is mainly provided by allocating sufficient resources during connection setup.

The responsibility to adhere to the negotiated bandwidth (average and peak) lies with the end-station. The network, however, may provide a policing function to guard against violation of the negotiated bandwidth. Packet-loss due to buffer overflows may be reduced by rate control mechanisms.

The header functionality in ATM is reduced to a basic minimum (five bytes). The main function of the header is to identify the switching at each node. Therefore, it is mostly composed of virtual connection and path identifiers. The reduced functionality of the header results in simplified processing at the switching nodes and reduced transmission overhead. Connection-based communication using small cells, however, requires that the cells are delivered in order of transmission. This places certain routing constraints on the network.

The payload of an ATM cell is 48 bytes. The small payload (and consequently small cell size) results in smaller internal buffers in the switch. The fixed cell size simplifies the switch fabric and the internal buffer management at each switching node. It also conducive to small delay and jitter characteristics.

Network delay characteristics are particularly important when real-time traffic is being handled. In an ATM-based network, information is assembled into cells at the source, transmitted over the network, and depacketized at the destination. For example, for voice transmission, a voice packet of 48 bytes at 8KHz sampling rate will incur a

	Constant Bit Rate (Class A)	Variable Bit Rate (Class B)	Connection Oriented Data Transfer (Class C)	Connectionless Data Transfer (Class D)
Timing between source and destination	required		not required	
Bit rate	Constant	Variable		
Connection mode	Connection Oriented			Connectionless

Figure 1: Services provided by ATM Adaptation Layer

6msec packetization delay. This packetization delay is proportional to the sampling rate and the packet size. The smaller the packet size, the smaller the packetization delay. Another component of the delay in the network for real time traffic is the depacketization delay, where a playout buffer may be used to even out the effect of jitter in the network. Jitter is caused by queueing delays due to resource contention. Contention between large packets causes greater jitter than contention between small packets. The queueing delay at the end-station and at the switching nodes is also reduced by using small packets.

In a broadband network, different applications requiring different services co-exist. These service requirements may vary dramatically. Hence, QoS characterization and support is an integral part of the ATM philosophy. Each virtual connection is established with a call setup procedure. At call setup time, in addition to establishing a circuit, the application QoS requirements for the connection are specified. The path for the connection is selected and resources are allocated, based on the requested QoS and the available network resources. The QoS parameters include: 1) bandwidth, 2) delay, 3) jitter, 4) cell loss probability, 5) call admission, etc. In a wide area network, the requested QoS also plays an import role on the tariff.

3 ATM

ATM defines four classes of services provided to the end user (Fig. 1):

1. **Class A.** Constant Bit Rate (CBR) circuit emulation (connection oriented).

2. **Class B.** Variable Bit Rate (VBR) with time synchronization between sender and receiver (connection oriented).

3. **Class C.** VBR connection oriented data service.

4. **Class D.** VBR connectionless data service.

Class A and B services may typically be used to support multimedia applications while Class C and D services may typically be used for packet data. This translates to strict delay and bandwidth guarantees for Class A, and more lenient bounds for Class B and C. Furthermore, coordinated resource allocation strategies are required to establish connections for the Class A, B, and C. Mechanisms must exist to manage and police connections in order to guarantee a negotiated QoS.

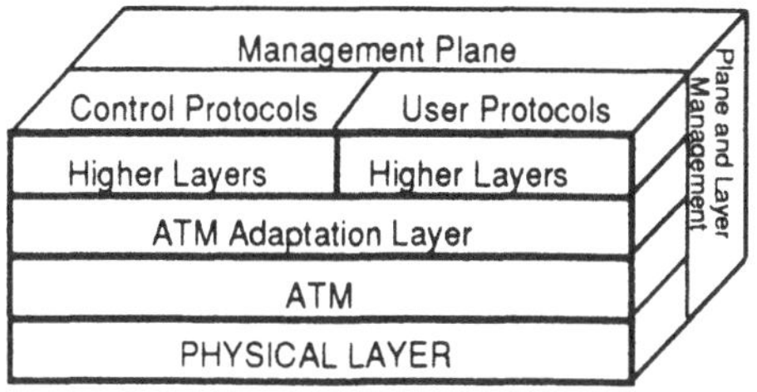

Figure 2: The layers of ATM networks.

The layered organization (Fig. 2) of ATM supports all the services defined above along with associated control and management functions. The management plane includes two types of functions: 1) Plane management, and 2) Layer management. The management functions that relate to the whole system are included in plane management. The layer management is, as the name suggests, based on a layered structure. It performs the management functions relating to resources and parameters residing in its protocol entities (e.g., meta-signalling). For each layer, the layer management handles the specific operations and maintenance (OAM) functions. The user plane provides for the transfer of application information including flow control or error recovery. The control plane is responsible for call control and connection control functions. These signalling functions are necessary to set up, supervise, and release a call or a connection.

3.1 The Physical Layer

The physical layer of ATM is subdivided into: 1) the Physical Medium Sublayer which supports pure bit functions, and 2) the Transmission Convergence sublayer wherein which ATM cells are converted to bits. The Transmission Convergence layer performs: 1) generation and recovery of the transmission frame, 2) transmission frame adaptation to adapt cell flow according to payload structure of the transmission system (e.g., SONET[8]) and to extract the cells upon reception, 3) cell delineation to identify cell boundaries, 4) HEC generation and checking, and 5) cell rate decoupling to insert idle cells to the payload capacity of the transmission system and remove idle cells upon reception.

CCITT defines physical layers at data rates of 155.520Mbps and 622.080Mbps using SDH-based (Synchronous Digital Hierarchy, CCITT Recommendation G.707) and cell-based interfaces. The ATM Forum has defined several physical layers at data rates of: 1) 44.735Mbps using DS-3, 2) 100Mbps using FDDI, 3) 155.52Mbps using SONET STS-3c, and 4) 149.760Mbps using Fiber Channel physical layers.

3.2 The ATM layer

The ATM Layer performs four main functions: 1) multiplexing and demultiplexing of cells from different connections onto the same cell stream, 2) translation of VCIs and VPIs (VCI and VPI are explained shortly) at switches or cross-connects (routing), 3) cell header removal and insertion, and 4) flow control.

ATM uses 53 byte cells for transmission and switching. The cell is composed of a 5 byte header and a 48 byte payload. The cell header format at the user network interface is shown in Fig. 3. The purpose of the GFC (Generic Flow Control) field is to alleviate

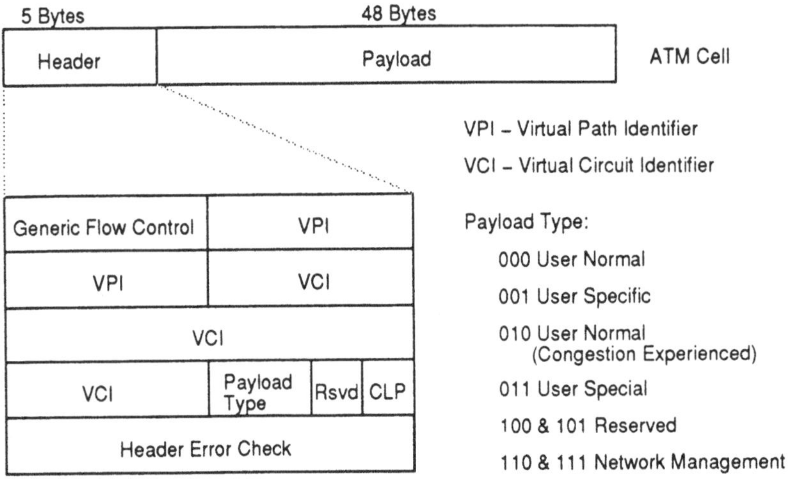

Figure 3: Format of the ATM cell header at user-network interface.

short-term overload conditions which may occur in the network. If an end-station is directly connected to an ATM switch, GFC may be used to throttle cell flow from the end-station. Alternately, in a shared media network, it may be used globally for media access control. The ATM network (network-network interface) does not provide any flow control mechanism. Therefore, there is no need for GFC within an ATM network. The GFC bits shown in Fig. 3 are used for VPI at a network-network interface.

3.2.1 Switching with Virtual Path and Channel Identifiers

ATM defines switching at two hierarchical levels: 1) Virtual Channel (VC) and 2) Virtual Path (VP), as defined in CCITT Recommendation I.113. The Virtual Channel concept describes a unidirectional transport of ATM cells associated by a common unique virtual channel identifier (VCI). The Virtual Path concept describes a unidirectional transport of cells belonging to virtual channels that are associated by a common virtual path identifier (VPI). A physical transmission link may carry many virtual paths, each of which may comprise many virtual channels. Thus, a virtual path may group several virtual channels to simplify switching. In general, VCIs and VPIs are used to identify transmission links. Switches identify cells by the VCI/VPI and translate incoming VCIs/VPIs into new VCIs/VPIs when forwarding cells to outgoing links. An example of a VC and VP switch is shown in Fig. 4.

Since VCIs and VPIs are associated with connections, they are assigned at connection setup time and freed at the termination of the connection. A set of connections (and thus the corresponding VCIs/VPIs) may be set up permanently for transport of network control and management data, as well as for connectionless datagram support.

3.3 The Adaptation Layer

The ATM Adaptation Layer (AAL) lies directly above the ATM layer and provides an interface to the service requirements of the higher layers. Higher layer data is carried in the ATM cell payload. The AAL is further divided into a Segmentation and Reassembly (SAR) sublayer and a Convergence sublayer (CS). The CS is service dependent and provides AAL service to applications.

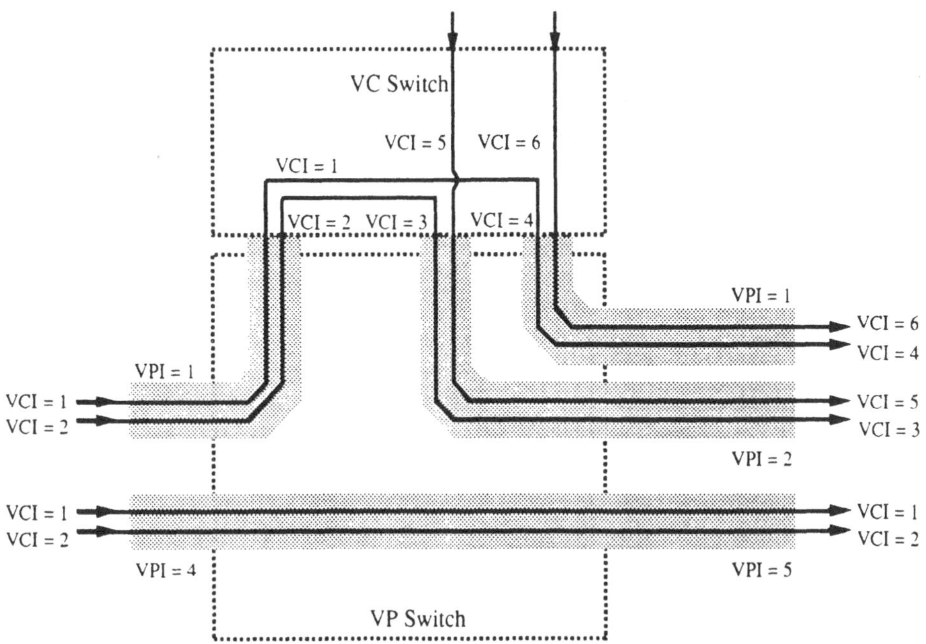

Figure 4: An example of switching using VC and VP.

Most data packets exceed the cell's small 48 byte payload (for example, a TCP/IP packet is at least 20 bytes of TCP and at least 20 bytes of IP header in addition to the data). Consequently, the majority of packets must be segmented and transmitted using multiple ATM cells. These cells are then reassembled upon reception. If the segmentation and reassembly is done in software, it will require significant processing within a cell time (e.g., at 155Mbps rate, a cell time is $2.6\mu sec$). The time available for processing is even smaller at higher rates. Consequently, the segmentation and reassembly function is typically performed in hardware.

Four ATM adaptation layer (AAL) types are defined to support different classes of service. A definition of each AAL type and their relation to a respective class of service is described below. We observe that ATM cells do not directly identify an AAL type. Consequently, it is the end-stations responsibility to associate the AAL type requested with the VCI/VPI of a connection. Similarly, switches need to associate QoS with VCI/VPI values to handle cells accordingly.

3.3.1 AAL Type 1

Normally, CBR Class A service will use AAL type 1 because it receives and delivers data at a constant bit rate. It also transfers timing information between source and destination, in particular, recovery of source frequency at the receiver is provided. The circuit emulation provided by AAL type 1, is believed to be an important feature of B-ISDN due to interoperability considerations. The SAR layer for AAL type 1 uses the first byte of the 48byte payload for a sequence number and sequence number protection bits. The sequence number is used to detect cell loss.

3.3.2 AAL Type 2

The AAL type 2 is commonly used to support Class B service (VBR). Like AAL type 1, it provides a source clock recovery function at the receiver. In addition to the user data, the AAL type 2 cell carries a sequence number, type of information, length, and CRC.

3.3.3 AAL Type 3/4

ATM adaptation layers 3 and 4 were recently combined into a common adaptation layer called AAL type 3/4. This adaptation layer is suitable for carrying both connection oriented (Class C) and connectionless data (Class D). Unlike AAL type 1 and 2, however, it does not provide any clock recovery. The segmentation and reassembly of AAL type 3/4 is shown in Fig. 5. The user PDU is encapsulated into a CS-PDU by the Convergence sublayer which may add an error checksum in addition to service access point and MAC address information. The CS-PDU is segmented and carried in the payload portion of ATM cells. As with AAL type 2, the sequence number, length, and CRC are used for segmentation. In addition, a Segment Type (beginning of message, continuation of message, end of message, and single segment message) message identifier exists to multiplex and demultiplex different PDUs over a single connection. This feature is useful when a single VCI is used to carry cells belonging to different messages.

3.3.4 AAL Type 5

AAL type 5 has been proposed as a simple and efficient adaptation layer mainly for Class D service (in particular IP datagrams). The user PDU is appended with a 2 byte length field and a 4 byte CRC. Then, it is transmitted in 48 byte blocks in each ATM cell. The payload type bit in the AAL type 5 type cell is used to indicate the end of the packet. At the receiving end, the cell payload is concatenated in the order of arrival until the end of packet is detected. The CRC and the length are checked to detect errors (lost or corrupted cells). Thus, AAL type 5 efficiently uses the bandwidth and provides a simple mechanism to identify data integrity. This is achieved by a simple SAR sublayer and by not allowing multiple messages to be multiplexed on the same virtual connection.

3.3.5 Services and Operations in AAL type 3/4 and 5

Two modes of services are provided by AAL type 3/4 and 5: 1) message mode service, and 2) streaming mode service. In message mode service, a complete PDU is transferred between the Adaptation layer and higher layers as a single unit. In streaming mode, the transfer of partial PDUs may be separated in time. This allows the Adaptation layer to expedite the transfer of PDUs to the higher layers by delivering them in stages.

For both services, an Assured and Non-Assured modes of operations are supported. In the Non-Assured mode, higher layers are informed of a lost or corrupted cell (optionally, a corrupted cell may be delivered). However, lost or corrupted cells will not be corrected by retransmission by adaptation layers. In principle, end-to-end flow control may be provided to a point-to-point connection. However, no flow control is provided for point-to-multipoint connections.

In the Assured mode of operation, data delivered to the receiver is guaranteed to be identical to the data sent by the sender. This is accomplished by retransmission of

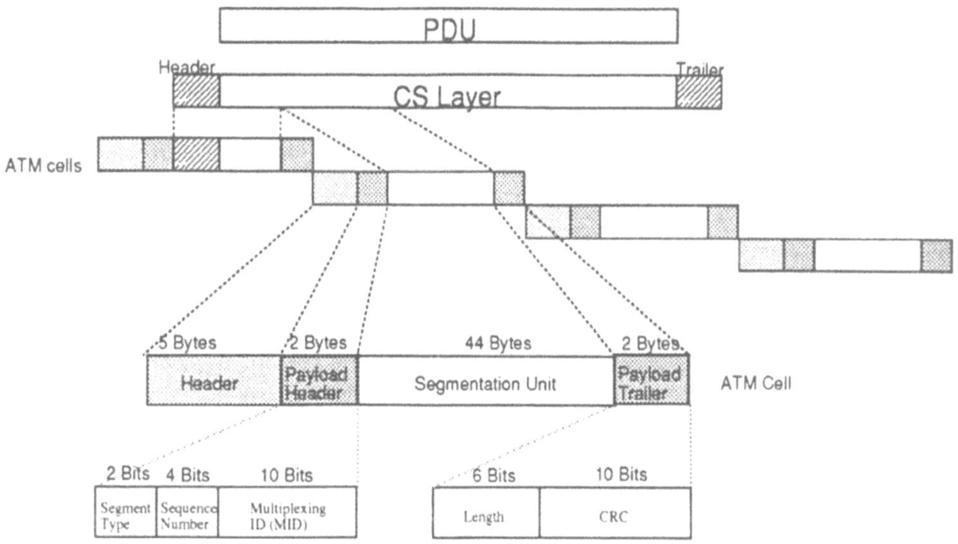

Figure 5: Adaptation of a user PDU using AAL type 3/4.

lost or corrupted cells. Flow control is provided as a mandatory feature and is restricted to point-to-point connections only.

In the next section, we describe DQDB: an ATM-like network which may be considered an introductory step towards ATM networking.

4 DQDB

Historically, in the early '80s, the need for a Metropolitan Area Network (MAN) standard was recognized. This led to the formation of the IEEE 802.6 committee. A number of proposal were submitted to and discussed by the committee: the interested reader will find summaries of these early proposals in [9, 10]. Late in 1987 the *dual bus* proposal by Telecom Australia, called QPSX, was agreed upon as the basis for a MAN standard[11]. The network became known as DQDB (Distributed Queue Dual Bus) and has since become the building block for the IEEE MAN. In July 1991, ten years after its inception, the IEEE 802.6 committee published its first MAN standard[12]. The document defines the necessary functions to support a connectionless service to a *logical link control* (LLC) sublayer in a manner consistent with other IEEE 802 LANs. The document also contains the framework for providing additional functions to support connection-oriented and isochronous services. Currently under study by the 802.6 committee, the latter functions are expected to appear in future additions (or in separate documents) of the 802.6 standard.

In the next few subsections, we will present a brief overview of the DQDB protocol.

4.1 The DQDB Network and the MAC Protocol

The DQDB network consists of two, counter-flowing, unidirectional slotted buses with a collection of nodes attached to both buses, see Fig. 6. The nodes can receive (resp. transmit) information from (resp. to) both buses thus providing the capability of full duplex communication between any group of nodes.

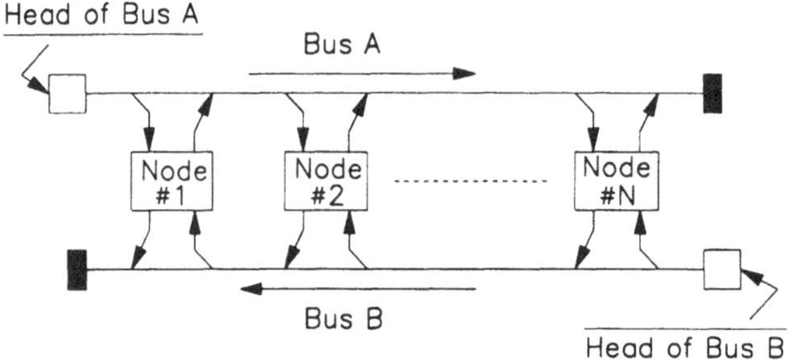

Figure 6: The DQDB network.

The nodes at the leading ends of their respective buses have the *head of the bus* (HOB) responsibility to continuously generate slots (or *cells*) of fixed duration. The slots "travel" along the buses and "drop off" at the end of them. Unlike ring networks, nodes are not required to remove information from the buses. For increased reliability, the two HOBs may be collocated, in which case the network will form a physical ring but still behave as a logical dual bus. In the event of a fault on the (physical) ring, the "gap" at the (collocated) HOBs may be closed by moving the HOB responsibility to nodes adjacent to the fault. Hence, the operation of the whole network can continue with no catastrophic consequences.

The DQDB MAC protocol (see below), although operating in a slotted environment, is not tied to any particular slot size. Over the years the slot size has varied from 256 octets to 45 octets to 69 octets. Finally, in order to gain maximum benefit from the emerging ATM standards, the slot size in DQDB was set at 53 octets. Of the 53 octets, 5 octets are designated for the *header* and 48 octets for the *payload.*

Nodes may access the DQDB buses in one of two modes. The *queue arbitrated* (QA) access mode, for the non-isochronous traffic, and the *pre-arbitrated* (PA) access mode, for isochronous traffic. The first octet in a slot constitutes the *Access Control Field* (ACF), see Fig. 7, which is primarily used by the nodes to coordinate their accessing rights to the network. The SL_TYPE bit designates the slot as a QA or PA slot. A node cannot transmit QA traffic in PA slots and vice versa. The *busy* bit indicates whether a slot carries information; no QA traffic can access a slot with the busy bit set. The *request* subfield is used by the nodes to reserve, from their upstream nodes[1], future QA empty slots; one request bit is used for each priority level allowed in the network but all connectionless communications must be done at the lowest priority level 0. The busy and request bits are originally set to 0 by the HOBs.

Like ATM, due to the limited size of a slot, QA traffic generated by a node will be grouped in *segments* (Fig. 5) big enough to fit in the payload portion of a slot. In the following subsection, we will present the QA mode access protocol for single priority traffic followed by a bandwidth allocation enhancement mechanism added to the protocol, the multi-priority case, and finally a discussion on additional services to be offered by the DQDB network.

[1]The terms "upstream" and "downstream" are with respect to the flow of traffic on the transmission channel.

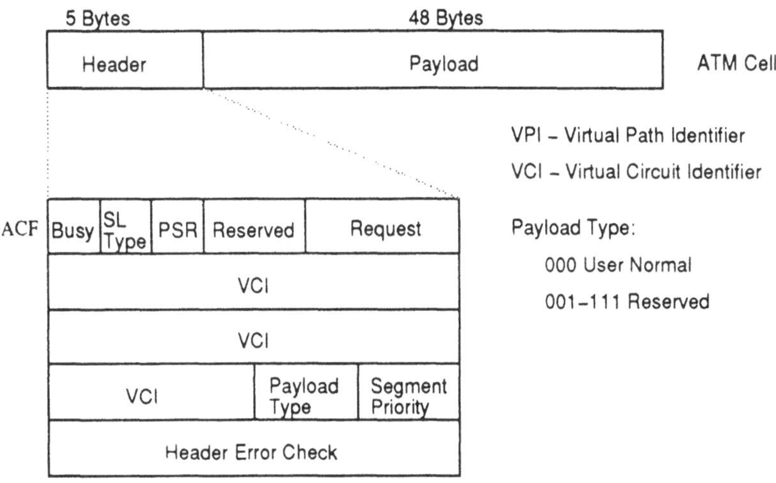

Figure 7: The DQDB slot format.

4.2 The Basic Distributed Queueing Algorithm

Without lack of generality, we consider segment transmissions on Bus A, see Fig. 6, to be called the *transmission* channel. Bus B will be called the *reservation* channel. The procedure described below applies unchanged for segment transmissions in Bus B with the roles of the transmission and reservation channels reversed.

Unlike other 802.X MAC protocols, DQDB nodes keep track of the congestion level in the network and adjust their segment transmission rate accordingly. The DQDB MAC protocol creates and maintains a virtual global queue of requests for segment transmissions on the transmission channel. A best effort is made for this queue to be served in first-in-first-out (FIFO) order. The global queue is realized in a distributed fashion by each node.

Fig. 8 depicts the state diagram for the DQDB MAC protocol. A node is either in the *idle* state (when it has no segment to transmit) or in the *countdown* state. When a node is in the idle state, it keeps track of the outstanding (segment transmission) requests from downstream nodes. In particular, the node uses a *request counter* (RQ_CTR) that is incremented for each request seen on the reservation channel and decremented for each empty slot seen on the transmission channel. The RQ_CTR never drops below zero—a feature that adds to the robustness of the protocol.

A node enters the countdown state when it obtains a segment to be transmitted. The contents of the RQ_CTR are copied to a second counter, called the *countdown counter* (CD_CTR), and the RQ_CTR resets to zero. The node also queues a request to be sent to upstream nodes by setting the request bit in the first request-free slot on the reservation channel. The countdown counter decrements, based on the empty slots seen on the transmission channel, until it reaches zero. The node transmits its segment in the first empty slot of the transmission channel following the instant that the CD_CTR indicates zero. While the segment waits for its transmission, the RQ_CTR counts all the new requests seen passing on the reservation channel. Following the segment transmission, if additional segments are waiting, the node repeats the whole

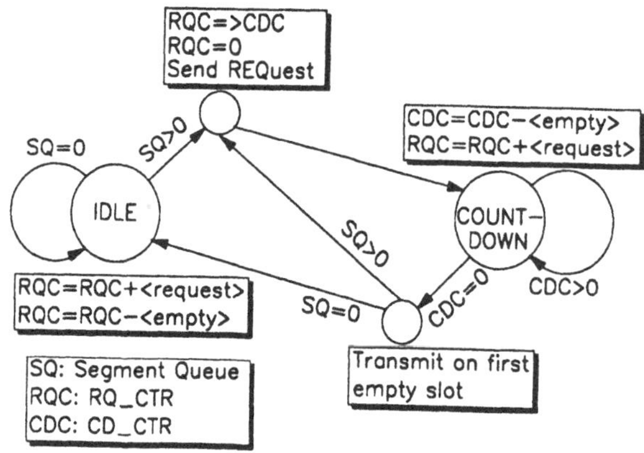

Figure 8: The DQDB MAC protocol state diagram.

procedure for the next available segment, otherwise the node returns to the idle state.

In a DQDB network, each node forms an "image" of the current state of the global queue and places its segments in the queue accordingly. Under ideal conditions (like zero propagation time and infinite request capacity) the queue image is identical in each node and segments are served in a true FIFO order. Ideal conditions, however, are physically unrealizable. Hence, the queue images at each node do not coincide and local FIFO service may not emulate the FIFO service of a true global queue. This condition gives rise to the highly publicized position dependent performance (or unfairness) characteristics of the network. We need to note however, that this characteristic is not unique to DQDB but exists for any linearly arranged network lacking a central control mechanism. In the next subsection, we present an enhancement to the basic DQDB MAC protocol that addresses the bandwidth allocation unfairness of DQDB in an overload situation.

4.3 The Bandwidth Balancing Mechanism

The request for transmission procedure used in DQDB becomes less effective as the distance-transmission speed product increases. As the distance-transmission speed product increases so does the number of slots in transit between nodes in the network. Consequently, reservation information becomes stale and the network cannot react fast enough to new requests for segment transmissions. In overload situations, the bandwidth allocation among the nodes is rather unpredictable.

To improve the bandwidth allocation scheme, a *bandwidth balancing* (BWB) *mechanism* (BWB) has been added to the basic DQDB MAC protocol[13, 14]. According to the BWB mechanism, a heavily loaded node will refrain from "hogging" the network by foregoing the opportunity to transmit even though the basic DQDB protocol would permit it to do so. By allowing empty slots to pass network bandwidth is freed for downstream neighbors whose explicit requests may not yet have propagated through the network. In particular, each node utilizes a BWB counter (BWB_CTR) that is incremented following each segment transmission by the node. When the BWB_CTR reaches the value *BWB_MOD* (which is a network parameter), the BWB_CTR resets to zero and the node refrains from transmitting in the next eligible empty slot.

The default value for *BWB_MOD* has been set to 8, which is a compromise between the reduction in maximum network utilization (because some of these extra empty slots may drop off unused) and the amount of time it takes to balance the bandwidth among the network nodes. The time for BWB to become effective increases with the distance-transmission speed product. For a network with N active nodes, all of which use the same value of BWB_MOD, the maximum network utilization η_{BWB} is given by

$$\eta_{BWB} = \frac{N \times BWM\_MOD}{1 + N \times BWB\_MOD}$$

4.4 The Priorities in DQDB

Although all connectionless data segments use priority level zero, future services may use the other two priority levels as well. Hence, for upward compatibility, the DQDB standard requires separate distributed queues for each of the three priority levels [12].

Without getting into specifics, the implementation of the priorities requires a separate pair of CD_CTR and RQ_CTR for each priority level on each transmission channel. While the node is in the idle state at priority level $i, 0 \leq i \leq 2$, the corresponding request counter RQ_CTR_i registers all the requests received at priority level i and above. While the node is in the countdown state at priority level i, RQ_CTR_i registers only the new requests of priority level i. Requests of any level j, $j > i$, are registered by the corresponding CD_CTR_j. The counters of all priority levels decrement similarly to the single priority case.

The priorities lose their effectiveness as the distance-transmission speed product increases or when the BWB mechanism is active.

4.5 Other services

As a MAN, the DQDB would be expected to accommodate traffic types beyond the connectionless data transfers. Although not fully specified and currently still under study, the DQDB standard contains the "hooks" for *connection-oriented* and *isochronous* transfer services.

The connection-oriented data services will share the same segmentation and reassembly mechanism (similar to ATM AAL type 3/4) that the connectionless data are using. Yet, some data fields in the header of a slot will be different (e.g., the VCI field would contain a different value for each virtual circuit connection; a connectionless data service uses the default VCI field of all 1's). The connection-oriented data segments will share the DQDB buses according to the same QA protocol previously defined.

The isochronous service would provide access of isochronous octets on a previously established connection on the DQDB network. The isochronous octets are transferred on PA slots, which the HOB pre-specifies by setting the SL_TYPE bit in the ACF of a slot, see Fig. 7. No QA data segments can access the slots with the SL_TYPE bit set. The PA slots are generated by the HOB periodically so that virtual circuits of guaranteed bandwidth can be formed. On occasion, buffering just prior or just after the receipt of the isochronous octets may be needed in order to accommodate instantaneous rate differences between the PA slot and isochronous octet generation. Contrary to the QA slots, several nodes can transmit in the same PA slot by accessing different octets. A node can access more than a single octet in a PA slot. Thus a wide range of isochronous transmission rates can flexibly be accommodated. The VCI value to be used for each isochronous connection and the offset position in a PA slot, that a node is allowed to access, are parts of the connection establishment phase which are not currently specified by the standard.

The above and other issues related to DQDB have generated a substantial amount of literature. A collection of DQDB research topics and a survey of associated literature is presented in [15].

5 DQDB and ATM

Both DQDB and ATM utilize 53 byte cells. Moreover, the 44 byte payload of a DQDB cell can be transparently transported across an ATM network using AAL type 3/4 cells. Hence, DQDB could provide a natural migration path to ATM.

Transparent interconnection between the networks, however, is not possible. This is due primarily to the fact that DQDB is based on distributed switching while ATM is based on centralized switching. In particular, unlike ATM, in DQDB there is no explicit connection acceptance phase and connectionless data may be transmitted without a prenegotiated connection identifier. Access to a DQDB network is unrestricted (other than by the fairness mechanisms in the MAC protocol), while access to ATM is subject to a service contract and traffic policing and regulation.

There are two hierarchically organized multiplexing levels for DQDB cells as opposed to three in ATM. In particular, DQDB cells are identified by a VCI field in the cell header and then by an MID field within the payload header. In ATM, although originally there was only a VCI field, this field was broken down into separate VPI and VCI fields in order to improve the ATM cell multiplexing flexibility[16, 17]. Moreover, since in DQDB there exist only one path which is shared by all the network nodes, the VCI field in DQDB is not really used for connectionless data traffic; the VCI field in DQDB has a default value of all 1's. In ATM the combined VPI+VCI field is 24 bits long as opposed to only 20 bits in DQDB. Both ATM (AAL type 3/4) and DQDB cells use a 10-bit MID field for an additional level of cell multiplexing.

Isochronous traffic is also treated differently by the two networks. In DQDB, isochronous data may be identified in the slot header and octets are preassigned to nodes on a request basis. In ATM, isochronous service is not defined in terms of fixed slots, but rather through the circuit emulation AAL type 1 traffic. Consequently, a gateway (protocol converter) may be used to join the two networks to compensate for their differences.

6 Discussion

When B-ISDN matures, it will carry applications and provide services which are difficult even to envisage today. Unquestionably, a major benefit gained by such a network would be the financial burden reduction in managing and operating a single network as opposed to an overlay of separate networks each for a different service provided. Still, even when the latter network integration becomes a reality, it will be highly disappointing if the voice, video, and data services provided over the network are not as good, if not better, than what is provided by existing dedicated networks for these services. It will be unfortunate and potentially disastrous for the future of B-ISDN if for the sake of its early release, an expensive integrated network solution is created which does not provide the same quality of service as do multiple independent networks, each one dedicated to a single service.

Before maturity is reached and the utmost possible benefit can be enjoyed, the evolution to the true B-ISDN based on ATM will undoubtedly undergo a number of development steps where the following issues will have to be gradually addressed and judiciously resolved.

ATM as a Data Service. Initially, it is expected that the bulk of traffic carried by an ATM network will be data traffic, e.g., LAN internetwork traffic. Hence, a major issue regarding ATM is the support for connectionless (datagram) traffic. Recall that ATM is a connection-oriented technology. All transfers are based on virtual connections established prior to the transfer of data. A large percentage of today's data network (both in WANs and in the LANs) is bursty short lived datagram traffic. The ATM connection-oriented paradigm does not work very well with datagram packets with arbitrary destinations. It is extremely inefficient to establish an end-to-end virtual connection every time a datagram packet is generated. One alternative is to build a map of *permanent* virtual circuits (PVCs) on top of the physical network topology to carry the connectionless traffic. Such a solution will eliminate the overhead of establishing end-to-end connections, but may introduces other complex issues. For example, criteria and algorithms for selecting these PVCs and procedures for allocating bandwidth to them.

Negotiating Quality of Service. As in any network, the full fledged B-ISDN network will be successful only when an end-station that gains access to the network "feels" as if it is the *only* beneficiary of the network's services. In order to attain such an *illusion*, a set of *Grade* and *Quality of Service* (GoS or QoS) objectives have to be established and quantified, the satisfaction of which will create the aforementioned illusion. The GoS relates to the rate of acceptance of new end-users to the network, similar to the call blocking probability in POTS ("plain old telephone service"), while a poor QoS could result in annoying clicks, glitches, echo, unintelligible voice, video-picture distortion, etc.

When an end-user requests the transport services of the network, the end-user provides the network with the QoS parameters required for the end-user's application. These parameters might include the end-to-end delay, jitter, the probability of cell loss, etc. The end-user also provides a *traffic descriptor* which is a set of parameters that specify the traffic characteristics of its application. Defining a traffic descriptor which will appropriately characterize an application is an important open issue. The traffic descriptor has to provide the maximum possible information about the application with the minimal set of parameters. Maximum information is needed in order to easily distinguish the multitude of applications that will access the ATM network. Given that decision algorithms (e.g., accept/reject a call, drop a cell) have to be executed very fast and, hence, be very simple, the traffic parameters have to be simple and minimal. Potential parameters in a traffic descriptor include the average transmission rate required by the application, the peak rate, the average length over which the peak rate is used, etc.

When a virtual connection is established across an ATM network it is made with a guaranteed QoS as negotiated with the user. These QoS characteristics are guaranteed so long as the user adheres to the negotiated usage of the connection. Should the user exceed the negotiated bandwidth, the QoS (e.g., delay and packet-loss) may degrade beyond guaranteed bounds. Yet, the traffic characterization may be particularly difficult for interactive-type applications which demonstrate significant variations in traffic behavior over a period of time (e.g., scientific visualization), while for other applications, the desired QoS is not known at call setup time. In the latter cases, a dynamic connection contract may be needed where the QoS parameters are renegotiate with the network as the demands dictate. Finally, the QoS parameters could be highly interrelated. For example, congestion can cause higher delay, jitter, and cell loss; while noisy links will only affect cell loss probability.

It follows from the previous discussion that a set of simple, sophisticated, flexible, and robust traffic controls have to be devised that will guarantee the GoS and QoS for all potential applications carried and services provided by the ATM network. Moreover, such traffic controls have to be active both before and after a call is accepted to the network. Below we discuss some of these controls.

When a new end-user wants to access the network, a decision has to be made through a *connection admission control* (CAC) whether or not access permission will be granted. The objective of the CAC is whether or not to accept the new end-user based on the availability of sufficient network resources (bandwidth, buffering capacity at affected intermediate network nodes, etc) that can provide the requested QoS to the new call without influencing the QoS of the already established connections. The CAC will work with the traffic descriptor and the QoS provided by the end-user; and based on its knowledge of the network status, the CAC will decide whether or not to accept the new call.

The acceptance by the network of a new end-user is a contract according to which the network commits to provide the requested QoS as specified by the traffic descriptor. Since an end-user can access the ATM network up to its maximum capacity, a *usage parameter control* (UPC) has to be provided to police and force the end-user to abide to the negotiated traffic descriptor. A malicious or unintentionally misbehaved end-user could easily degrade the QoS of established connections if it is allowed to freely violate its declared traffic specifications. Hence, a useful traffic descriptor has to be both easily observable and controllable. A *cell loss priority* (CLP) bit exists in the ATM header (see Fig. 3), that the network can set if the cell violates its traffic descriptor. During adverse network conditions, the network may drop cells with the CLP bit set. The policing mechanism has not yet been defined.

Congestion Control. Even when all the end-users accessing the network are well behaved, unexpected traffic fluctuations may give rise to congestion. In such a situation, due to an overload of the communication resources, the network cannot reliably provide the agreed upon QoS to the traffic streams (or applications) involved. Traditionally, congestion and flow control mechanisms would isolate the congested resource, redirect the traffic around it, and control the sources of the congestion. However, in an ATM network with its large distance-transmission speed product, hundreds or thousands of cells may be in transit towards the congested site at the instant that the congestion arises (and many more cells may enter the network) long before any action against the congestion becomes effective. Redirection is difficult since the cell order of transmission must be maintained. Such a situation puts excessive pressure on the buffers of the network nodes. The QoS parameters of the applications involved could very easily be violated in such a scenario. Hence, it is expected that congestion control should be preventive rather than reactive. That is, the congestion control mechanism should be activated before even congestion arises! Potential CACs and UPCs in ATM networks should incorporate sufficient safety margins for congestion controls to be effective.

Before acceptable and flexible traffic control procedures are devised, "first generation" ATM networks may rely on simple "peak bit rate" traffic controls (like in circuit switched networks). In this case, the efficiency of the network may be low. Nevertheless, valuable knowledge will be gained by studying the characteristics of the traffic generated by the various applications transported on such an ATM network. This knowledge will assist in better understanding the still unknown dynamic behavior of an ATM network and in the subsequent development of more sophisticated and efficient multiplexing and control procedures. At the same time, though, due to the traffic characteristics of future and still unknown applications, such procedures have to be robust with respect to the specifics of each application.

The hierarchical multiplexing capability of ATM cells will potentially decrease the complexity of integrating various service types. Applications with similar characteristics and QoS requirements could be aggregated into groups and separate VPs could be dedicated for each such group. Traffic control procedures could then be devised to address QoS issues within each VP separately (i.e., type of application), which is a much simpler task. The progressive incorporation of more effective control procedures will substantially increase the efficiency of the network and the range of applications that the ATM network will carry.

The Scalability of ATM. Since ATM is intended to be *The Network* for many many years to come, enough flexibility should be embedded within the network so as to easily accommodate future communications technology advances.

ATM is being defined across a large range of speeds (e.g, 100Mbps to Gbps). However, an ATM network designed to run at 600 Mbps will not necessarily run at, say, 2 Gbps. Although these link speeds may be achievable, there are many other components in the network which have to be modified to support the higher bit rate. First of all, the end-user adapter (including protocol processing and device driver) must be capable of feeding the link at 2 Gbps. Second, the Line Interface Adapters to the switch fabric, responsible for scheduling cells on appropriate links through the switch fabric, must be able to perform the virtual circuit routing at Gbps rates. Finally, the switch fabric itself needs to run at Gbps rates. Should all these factors be taken care of, it will be an expensive solution for the 600Mbps version, and obsolete when tens of Gbps are required. Hence, the physical ATM networks installed in the '90s will not be the end-all network solutions. Although the ATM philosophy can be extended to higher rates, existing networks will have to be replaced to support these higher rates.

It is implicit in the connection-oriented direction of ATM networks that the call establishment and release phases are practically negligible in length as compared to the actual information transmission phase. With the introduction in the future of more advanced communications technologies, the information transmission phase may shrink for several of the broadband applications (but not for all, e.g., real-time transmissions of voice or video), while the call establishment and release phases have an absolute minimum which is dictated by the physical size of the network. Potentially fewer applications could be served by the connection-oriented communication as the ratio between call establishment and data transfer increases. Furthermore, algorithms for resource allocation, QoS determination, etc., may need to change when moving to higher speeds. Hence, the ATM philosophy, if not implemented wisely, may not scale gracefully to higher speeds.

7 Summary

ATM, the internationally agreed upon technique for transmission, multiplexing and switching in a broadband network, is designed to support the integration of high quality voice, video, and high speed data traffic. Although ATM is a transfer mode suitable for implementing broadband networks (WAN, MAN, and LAN) it is not a network architecture in and of itself. Interestingly, the IEEE 802.6 DQDB Standard whose origin is unrelated to ATM will be the introductory step to ATM and B-ISDN[4].

Although much work is underway in the definition, design, and development of ATM, many important components of the ATM network have not been solved. DQDB can be viewed as a specific ATM network implementation which addresses some of these issues within a more constrained environment. Before maturity is reached and the utmost possible benefit can be enjoyed, the evolution to the true B-ISDN based on ATM will undoubtedly undergo a number of development steps where the many issues will have to be gradually addressed and judiciously resolved. These issues include:

1. Support for connectionless data traffic.
2. Methods for characterizing application requirements in order to specify QoS.
3. Guaranteed QoS with preventive rather than reactive congestion control.
4. Scalability to higher speeds and larger number of users.

When B-ISDN matures, it will carry applications and provide services which are difficult even to envisage today. Unquestionably, a major benefit gained by such a network would be the financial burden reduction in managing and operating a single network as opposed to an overlay of separate networks each for a different service provided. Still, even when the latter network integration becomes a reality, it will be highly disappointing if the voice, video, and data services provided over the network are not as good, if not better, than what is provided by existing dedicated networks for these services. It will be unfortunate and potentially disastrous for the future of B-ISDN if for the sake of its early release, an expensive integrated network solution is created which does not provide the same quality of service as do multiple independent networks, each one dedicated to a single service.

The new cell-based technologies of DQDB and ATM are expected to revolutionize communications. Without having to wait until the next millennium for all the problems to be solved, we may prepare ourselves with a step-by-step introduction to the new services. Such a process will familiarize ourselves with the new concept of broadband networking and will better prepare us to solve the new challenges imposed by such a universal communications environment.

References

[1] E.D. Sykas, K.M. Vlakos, and M.J. Hillyard. Overview of ATM networks: functions and procedures. *Computer Communications*, 14(10), December 1991.

[2] Martin de Prycker. *Asynchronous Transfer Mode: Solution for Broadband ISDN.* Ellis Horwood, 1991.

[3] R. Handel and M. N. Huber. *Integrated Broadband Networks: An introduction to ATM-Based Networks.* Addison Wesley, 1991.

[4] M. de Prycker. ATM Technology: A Backbone for High Speed Computer Networking. *Comp. Netw. and ISDN Syst.*, 25(4-5):357–362, November 1992.

[5] Bellcore Technical Reference. *Generic system requirements in support of Switched Multi-megabit Data Service*, May 1991. TR-TSV-000772.

[6] F. R. Dix, M. Kelly, and R. W. Klessing. Access to a public Switched Multi-megabit Data Service offering. *Comp. Commun. Rev.*, 20(3):46–61, July 1990.

[7] W. Stallings. *Handbook of Computer Communications Standards*, volume 2. MacMillan Book, 1990.

[8] ANSI Standard T1.105-1988. *SONET Optical Interface Rates and Formats*, 1988.

[9] R. W. Klessing. Overview of Metropolitan Area Networks. *IEEE Communications Magazine*, 24(1):9–15, January 1986.

[10] J. F. Mollenauer. Standards for Metropolotan Area Networks. *IEEE Communications Magazine*, 26(4):15–19, April 1988.

[11] R. M. Newman, Z. L. Budrikis, and J. L. Hullet. The QPSX MAN. *IEEE Communications Magazine*, 26(4):20–28, April 1988.

[12] IEEE Std 802.6-1990. *IEEE Standards for Local and Metropolitan Area Networks: Distributed Queue Dual Bus (DQDB) Subnetwork of a Metropolitan Area Network (MAN)*, July 1991.

[13] E. L. Hahne, A. K. Choudhury, and N. F. Maxemchuck. DQDB networks with and without bandwidth balancing. *IEEE Trans. on Commun.*, 40(7):1192–1204, July 1992.

[14] C. Bisdikian. A Performance Analysis of the IEEE 802.6 (DQDB) Subnetworks with the Bandwidth Balancing Mechanism. *Computer Networks and ISDN Systems*, 24:367–385, 1992.

[15] B. Mukherjee and C. Bisdikian. A Journey Through the DQDB Literature. *IBM Res. Rep., RC 17016*, July 1991. to appear in Performance Evaluation, Dec. 1992.

[16] S. Ohta, K. Sato, and I. Tokizawa. A Dynamically Controllable ATM Transport Network Based on Virtual Path Concept. *Proc. GLOBECOM'88*, pages 1272–1276, November 1988.

[17] Y. Sato and K. I. Sato. Virtual Path and Link Capacity Design for ATM Networks. *IEEE J. Sel. Areas Commun.*, 9(1):104–11, January 1991.

ON TRANSPORT SYSTEMS FOR ATM NETWORKS

M. Zitterbart[1], A.N. Tantawy[1], B. Stiller[2], T. Braun[2]

[1]IBM Research Division
Thomas J. Watson Research Center
P.O. Box 704, Yorktown Heights, NY 10598, USA

[2]University of Karlsruhe
Institut of Telematics
Kaiserstr. 12, 7500 Karlsruhe, Germany

ABSTRACT

Emerging ATM networks accompanied by an increasing variety of application service requirements may change future transport subsystem design in order to provide enhanced service characteristics and improved performance. System designs integrating proper architectures, services and protocols are increasingly required. This paper gives a survey of some research work dealing with transport subsystem issues for ATM network. A framework for an ATM transport system is presented providing an enhanced service interface and supporting application-driven flexible protocol configurations. The framework uses a non-hierarchical, fine granular function-based decomposition of the communication task in order to efficiently provide these features.

1. INTRODUCTION

Architectures, services and protocols of transport systems may have to undergo noteworthy changes in order to cope with the characteristics of future networks and applications. New types of services need to be provided for emerging applications, such as distributed multimedia applications. In addition, highly efficient transport systems are required in the future to provide the performance of high speed cell-based networks to applications. Generally, this situation is widely recognized; several new protocols have been developed (e.g., [1], [2], [3], [4]) and various approaches towards efficient implementations have been presented (e.g., [5], [6], [7], [8]). However, there are not many promising approaches addressing the different issues and presenting integrated solutions. An example is the Esprit Project OSI 95 [9], in which high performance OSI protocols and services for multimedia support on HSLANs and B-ISDN have been investigated over the last couple of years.

Asynchronous Transfer Mode Networks, Edited by Y. Viniotis
and R.O. Onvural, Plenum Press, New York, 1993

Standard organizations, such as ISO, have realized the need for changes in order to satisfy current technology and applications. ISO has recently started a new work item on Guidelines for Enhanced Communication Functions and Facilities for the Lower Layers [10]. It comprises issues on transport service, architecture and modeling as well as efficiency and performance. In the ISO/IEC JTC1/SC6 meeting in San Diego, in July 1992, various proposals for this new work item have been presented. Due to the variety of alternative approaches, it cannot be foreseen at this point in time which way this effort will go.

The appearance of ATM networks, such as DQDB [11], SMDS [12], and BISDN [13], may have a high impact on the design of future transport systems. Due to some of their inherent characteristics they seem to provide a sound basis for emerging communication systems. ATM networks are becoming increasingly popular, not only for metropolitan and wide area networks but also in the local area [14]. However, transport systems consist of more than the network itself; they comprise higher layer protocols that provide end-to-end services to transport users across network boundaries. Therefore, higher layer protocols, services and the overall transport system architecture have to be considered in an integrated way -including issues on efficient implementations- to provide proper solutions for this rapidly changing communication environment. In this paper, issues and concepts related to transport systems over ATM networks are discussed and a framework for an adequate transport system is presented. Multiple aspects, such as transport services, protocol architectures and implementation efficiency, are addressed in an integrated manner.

Section 2 of this paper briefly summarizes some characteristics of ATM networks that may have an impact on future transport systems designs. A survey of some recently presented research work targeting issues related to transport system design and implementation for ATM networks is given in section 3. Section 4 presents a framework for a flexible high performance transport system. Finally, section 5 concludes the paper.

2. ATM NETWORK CHARACTERISTICS AFFECTING TRANSPORT SYSTEMS

2.1 Data Transfer Units

The most prominent difference between ATM networks and other existing networks and services (e.g., the Frame Relay [15]) is the use of small, fixed size transfer units, the cells. A *cell* has a length of 53 bytes consisting of a 48 byte payload and 5 bytes of control information [16]. Cells are routed through the network based on a label in the cell header. These relatively small data units bring up some issues in the implementation of suitable transport subsystems. Since cells may arrive at a very high rate (every 680 ns in a 622 Mbps ATM) high processing power is required in nodes that have to process functions on a cell-by-cell basis. This burden is moved outside the network to the end nodes in order to provide to provide for fast cell switching inside the network. As a result, end nodes have to cope with time critical cell-related operations, such as segmentation of larger higher layer data units into cells or reassembling cells to the original data unit. Moreover, error and congestion control at the cell level may be required (e.g., bibref refid= BaeS91.).

2.2 Multiple Network Services

Another salient feature of most ATM-based networks is the provision of *multiple types of services* to the higher layers of a transport system. Traditional networks, such as 802 LANs [17], provide only a single connectionless data service at the MAC service interface. FDDI [18] provides two services, an asynchronous and a synchronous service. A wider variety of services is highly required by the diversity of available and emerging applications. BISDN provides multiple services, including an isochronous service. Therefore, they provide a sound basis for a variety of traffic types needed to serve emerging multimedia applications which may comprise voice, video, and data traffic. However, typically the network does not directly interface to applications. Additional protocols have to be passed before a transport service is provided. Although some services offered by ATM networks may appear similar to traditional transport services, they are only provided at the borders of ATM networks. To interconnect with other networks, a network layer protocol is still needed to provide network wide routing capabilities. As a result, in a scenario using current standard protocols, the service capabilities of ATM networks may be buried underneath a TCP/IP [19] protocol stack. Consequently, changes in the services and protocols of transport systems are required to reflect the new types of network services at the transport service interface.

2.3 BISDN Protocol Architecture

The protocol architectures for ATM networks follow the layered philosophy of the OSI Reference Model [20] by defining several additional sublayers. They comprise some global features including link-by-link cell transfer common to all services and service-specific adaption protocols [21], [22]. However, the details are different as shown in Figure 1 for the BISDN ATM protocol architecture.

The ATM BISDN protocol architecture consists of two layers on top of the physical layer: the *ATM layer* which provides cell transfer capabilities and the *ATM Adaption Layer (AAL)* which supports higher layer functions. The ATM layer operates on cell headers, the AAL layer on the cell information field [16]. The AAL layer is further subdivided into two sublayers, the segmentation and reassembly (SAR) layer and the convergence layer (cf. Figure 1). AAL protocols are service dependent; currently four types of services to upper layers are specified [13], Class A to D. The service types are mainly distinguished according to the time relation between source and destination, variable or constant bit rate, and the selected connection mode. Due to the multiservice appearance of BISDN a one-to-one mapping of its architecture on the OSI Model is difficult to derive [23]. Some protocols actually fit under the MAC service (e.g., Class D), others are closer to typical transport services (e.g., Class A). Furthermore, the BISDN architecture defines

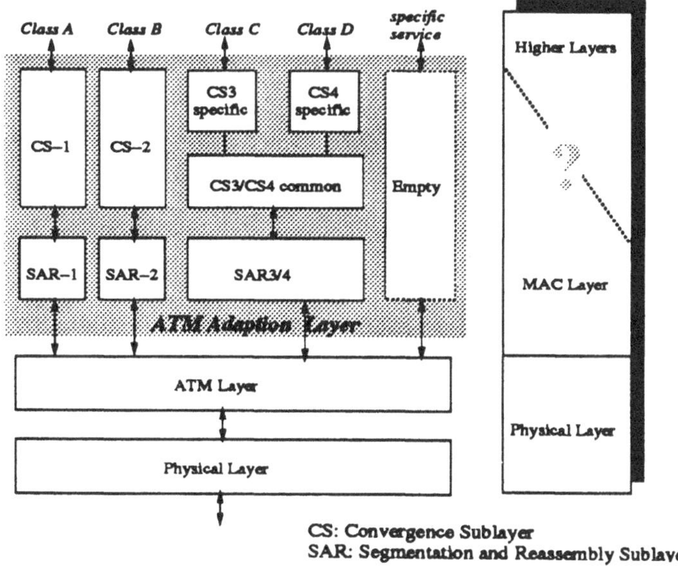

Figure 1. BISDN ATM protocol architecture

multiple planes: user plane, control plane and management plane. This especially emphasizes out-of-band signalling for connection management purposes.

3. SURVEY OF CURRENT RESEARCH ACTIVITIES

Cell-based ATM networks increasingly influence the communications world. So far, discussions around ATM mainly focused on network related aspects or on dedicated higher layer protocol mechanisms. However, higher layer issues including transport and network protocol design, transport service provision and architectural integration have to be revised carefully. Simple patches to existing solutions do not seem to be sufficient, for instance, for efficiency and flexibility reasons. This section presents a survey of some current research activities on transport subsystems closely coupled with ATM networks.

3.1 ATM Network Services and Current Transport Layer Services

The enhanced service interfaces of ATM subnetworks result in a gap between the service provided by the subnetwork and the service available at the transport service interface. The same ATM network can provide connectionless, connection oriented, isochronous services, etc. However, this variety is currently not reflected at the transport layer service interface. Commonly, transport protocols provide a single service, typically either a connection oriented service, e.g., TCP or a connectionless service, e.g., UDP. An overview of the advanced BISDN service classes and their requirements is given in [13][24]. The AAL service interface to higher layers is described in [25]. For example, for AAL type 1 constant length service data units with a constant inter-arrival time are defined. AAL type 3/4 provides message and streaming mode services with several selectable parameters. In [26], parameters related to the traffic characteristics in BISDN networks are described including peak cell rate, average cell rate, burstiness, etc.

In the ISO world, the characteristics of a transport connection can be described using a small number of Quality of Service (QoS) parameters [27]. Considering the data transfer phase, for example, throughput and transit delay as well as a residual error rate and a transfer failure probability can be specified. However, those parameters do not efficiently express future service characteristics, e.g., the delay jitter for continuous media or the response time for client/server applications. The available QoS parameters also do not support the definition of guaranteed values, e.g., a minimal guaranteed throughput or a maximum guaranteed delay jitter for a connection as it may be required for continuous media applications. Moreover, currently the QoS cannot be changed during the lifetime of a transport connection. Compared to other existing transport services, the ISO transport service is quite advanced. For example, the TCP protocol does not provide any quantitative QoS parameters at its service interface. Even recent protocol developments, such as the XTP protocol, lack the availability of a clearly specified enhanced service interface. In XTP several qualitative protocol options can be selected via a bitmap field in the packet (e.g., disabling the data checksum). However, the definition of quantitative parameters is not supported. Clearly, new transport services are needed in order to better utilize ATM network service capabilities and, thus, to satisfy a higher diversity of application requirements.

Several approaches towards *enhanced transport service interfaces* for communication platforms on ATM networks have been recently reported. Generally, a trend to finer granularity in transport service interfaces can be observed (e.g., [28], [29]) providing a larger set of distinctive service parameters and/or service classes. More service parameters are made visible at the transport service to enable fine granular, application-tailored service specifications.

Within the BERKOM project [30], a set of transport services is defined including a connectionless service, a connection oriented service and an advanced transport service. Moreover, extensions to QoS parameters (e.g., maximum delay variation, burstiness, etc.) are proposed [31]. In [32], a somewhat dynamic service interface providing advanced connectionless and connection-oriented types of services is presented. Compared with traditional services, they provide additional parameters, e.g., to select certain flow control mechanisms or to define QoS parameters, such as connection duration and delay jitter. The transport service being defined within project OSI 95 presently consists of five different types ranging from an enhanced connection-mode service to a basic (unacknowledged) connection-less service [33]. To achieve guaranteed quantitative QoS parameters, the concept of compulsory QoS parameters is introduced [34]. The transport system guarantees this compulsory value during the lifetime of a connection and initiates connection release in case it cannot be achieved longer. An example of a compulsory value could be the minimal requested throughput needed by an application. In the Flexible Communication Subsystem (FCSS) [28], discussed in more detail in section 4, different types of services are defined independently from the notion of connections. The selection of various qualitative parameters, such as ordered delivery of data or error tolerance, is provided for the user to specify its communication requirements. In addition, a set of quantitative parameters is given to enable transport users to request enhanced services. The project OSI 95 and FCSS also highlight the importance of *flexibility* and *dynamics* of QoS selection and control as they are increasingly required by forthcoming applications, such as distributed multimedia applications. In this context, negotiation and re-negotiation of services are considered important, even if they go along with a transparent re-establishment of a connection under the transport service level [29], [34]. This may be necessary in an ATM network since certain characteristics are statically bound with a virtual path/channel. Multimedia applications comprise multiple data streams, for which a synchronization services may be are required. Therefore, the orchestration service for coordination of multiple transport streams can be used [35]. In [28] this service can be explicitly selected by the user and is then implemented by a session manager.

3.2 Transport Subsystem Architecture

Several early transport system architectures especially considering ATM networks have been developed by mainly mapping currently used network and transport layer protocols on top of ATM protocol architectures, e.g., above the ATM Adaption Layer in BISDN [30], [31], [32], [36]. Most frequently, protocol stacks around the OSI TP4 protocol [37], the TCP protocol or the XTP protocol are selected. For example, in [30] an OSI protocol stack consisting of the CLNP network protocol [38] and the TP4 transport protocol is directly mapped on top of the adaption layer (TCP/IP could be used alternatively) to form the BERKOM communication platform for multimedia applications. The BERKOM-FINE protocol architecture [32] represents an alternative design providing a family of adequate high speed transport protocols for ATM networks.

Directly integrating protocol architectures of emerging networks in the architecture of the OSI Reference Model is common practice and, in principle, may look like the right way to go. However, at a certain point in time, those architectures should be carefully reviewed since too many patches do not necessarily make them any better. The overall architecture becomes increasingly complex, e.g., due to the development of additional sublayers (e.g., in case of ATM networks). This seems to be somewhat conflicting with the very high performance requirements of future communication systems.

For that reason, among others, different subsystem architectures overcoming the shortcomings of layered architectures and their protocols and being more adequate for emerging communication requirements have been recently proposed. The horizontally oriented protocol structure (HOPS) is introduced in [39]. The protocol architecture is decomposed into functions, instead of layers, thus, decreasing the layer inherent overhead and providing a higher flexibility. The Modular Communication Machine presented in [40] also abandons layering and favors a modular decomposition of the communication task into functional components which then may be efficiently implemented via dedicated hardware. The FCSS [28] uses protocol functions and protocol information bases as atomic building blocks and views the subsystem as component without internal layering. The x-Kernel [41] provides an implementation environment for communication protocols [42]. The ADAPTIVE project uses the x-Kernel approach as an implementation environment for a flexible and fine granular transport system [43]. Flexibility is also addressed in [44]; the granularity of flexible units is based on protocol entities.

3.3 Protocol Mechanisms and Architecture

Clearly, current protocols, especially certain protocol mechanisms, are not adequate for the use in high speed ATM networks. Particularly the high bandwidth-delay product influences the design of proper protocol mechanisms. For example, window-based flow control is not appropriate for such networks [33]. In [45] error control and connection management mechanisms for ATM networks are discussed. Several error control mechanisms are compared in [46] and a connection management mechanism using synchronized clocks is presented in [47]. Protocol mechanisms are not further discussed within this paper, however, the framework presented in section 4 is designed to integrate alternative protocol mechanisms for the same protocol function in order to allow optimized protocol configuration based on information about actual network characteristics and application requirements.

Another important topic in this context is a proper protocol architecture for future transport systems. The enhanced transport service requirements could be served by a single super protocol or by a set of protocols each appropriate for a specific service [48]. The current trend is clearly oriented towards the support of different protocol sets each associated with certain service capabilities; a single transport subsystem protocol covering all different needs is not considered as a practical solution. Going one step further, a protocol can generally be viewed as a set of functions which are implemented by specific mechanisms, e.g., the acknowledgement function using a selective mechanism, as described in section 4 (cf. [28]). Such a design allows flexible protocol configurations on one hand and efficient implementations on the other hand. Considering implementation efficiency during the protocol design phase is a very important issue since current transport subsystems do not perform nearly as well as available networks [49]. A prominent example for such an approach is the XTP protocol, which was designed with a VLSI implementation in mind.

4. FRAMEWORK FOR AN ATM TRANSPORT SYSTEM

Within this section, a framework is described for a future transport system adequate for the requirements of emerging applications over ATM networks. It integrates service, protocol and architectural issues as well as implementation issues in a single transport system design. The basic framework architecture is illustrated in Figure 2 and the main concepts are briefly outlined in the following subsections.

4.1 Transport Subsystem Architecture

The ATM transport subsystem framework is designed around a configurable set of fine granular atomic *building blocks* representing protocol functions and protocol information

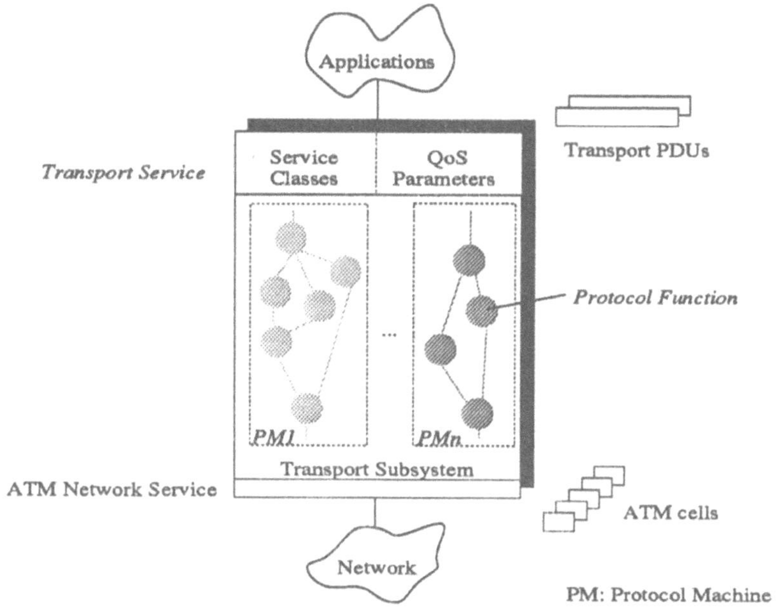

Figure 2. Framework Architecture of an ATM Transport System

bases typically located in the OSI layers 2b to 4. The building blocks are combined to *protocol machines* according to the service requirements specified at the enhanced fine granular transport service interface. Protocol functions are not subdivided any further; they are comparable to the concept of application service elements used at the OSI Application Layer [50]. Strictly hierarchical concepts, such as the layered OSI Model, are avoided for two reasons. First a high degree of service flexibility is anticipated to be increasingly important to forthcoming applications. Due to the finer granularity, the service offered by the communication subsystem can be adjusted more precisely to individual application requirements, omitting any unnecessary functions and, thus, potentially leading to more efficient implementations. Second, layered concepts normally impose a high overhead due to some of their inherent characteristics, such as inter-layer communication, layer transparency, etc. [40]. Furthermore, the concept of fine granular building blocks does not imply the serialization of function processing and, thus, provides a sound basis for efficient multiprocessor implementations. Concurrent processing of independent functions on the same data unit is supported to increase implementation performance. The concept of functional building blocks can also be applied to existing standard protocols, as, e.g., presented in [5] and [51].

The transport subsystem is architected in a way that allows for a direct mapping on top of the ATM Adaption layer. Protocol machines can be configured to fit between the available AAL service and the requested transport service. The transport system is also capable to serve raw ATM cell streams as they may appear if both the SAR and the CS sublayers of the AAL layer are empty (cf. Figure 1). This is expected to be an efficient mapping since it allows the tools of the transport system to completely decide on the protocol functionality needed. Internally in the transport subsystem, no further sublayering is considered.

4.2 Enhanced Transport Service Interface

A salient feature of the presented framework for an ATM transport subsystem is its service interface which supports application-driven configuration of efficient protocol machines tailored to the individual needs of the different applications. A detailed description of this service interface which is part of the Flexible Communication Subsystem is presented in [28] and [52]. This *enhanced service interface* is not based on the traditionally used notion of connections. The reason for that is the increasingly weak delimitation between the notions connection and connectionless and the believe, that applications requirements can be expressed more precisely using a set of qualitative and quantitative service parameters. The transport user does not care whether an acknowledged or an unacknowledged connectionless service or a connection-oriented service is provided. It is concerned about the way the data units are delivered at the service interface. The qualitative service criteria defined in the FCSS service interface precisely provide this capability.

Qualitative criteria include, among others, parameters to specify the requested flow of data at the service interface. For example, the ordered delivery of data can be requested or a semantics for a guaranteed group delivery can be specified (e.g., at least one, etc.). Furthermore, the user can define whether data loss, data replication, or data corruption can be accepted. If this is the case, a threshold value can be given using the corresponding quantitative parameters. Based on this knowledge, the transport subsystem can, e.g., decide whether an acknowledgement or a sequencing function is needed; for application this procedure should be completely transparent.

Quantitative criteria include parameters, such as throughput, delay, response time, rate, jitter, as well as data corruption and data loss thresholds. The quantitative parameters are specified in more detail using three different value levels: the threshold, the average and the useable values. For example, in case of the throughput parameter, the threshold value represents the minimum throughput with which the application can operate and the usable value specifies the maximum throughput that can be utilized by that application. The average value applies to the requirements over a period of time. However, most traditional applications might request only a threshold value, due to the lack of more specific knowledge about their communication requirements.

The quantitative and qualitative criteria provide a comparable fine granular transport service interface. In addition to this variety of service parameters, a set of *pre-defined service classes* is supported. The application can directly select qualitative and quantitative parameters or simply use the pre-defined service classes. Additionally, both levels can be combined, e.g., to refine the requirements on a specific service class. Service classes may, e.g., be used by applications that cannot specify their communication requirements in detail. Four pre-defined service classes are distinguished based on their support of isochronism, real time and reliability, namely, an unreliable real time class, a reliable real time class, an unreliable non-real time class, and a reliable non-real time class. Current applications may typically use the service class interface. However, it is anticipated that emerging applications will have an increasing knowledge about their specific requirements and, thus, use the finer granular parameter-based interface level. Those applications that cannot even specify a service class are provided with a very basic communication service only.

The pre-defined four service classes are appropriate to efficiently support current and emerging applications. To be widely accepted, an easy mapping of important current applications to service classes should be implemented. Therefore, we discuss here the *remote procedure call (RPC)* as a demanding application since it is the predominant communication paradigm applied in emerging distributed systems. It is also believed that efficient transactions are of particular importance for multimedia applications [35]. The required communication service can be characterized as follows. The error rate should be as low as possible. In particular, data corruption is not acceptable. Error tolerance in case of loss or replication of data may be acceptable, since most of the current RPC systems still work with internal acknowledgements at the RPC level. Nevertheless, from the performance point of view, an implementation of these mechanisms at the communication subsystem is preferable. Ordered delivery of data should be guaranteed at the RPC interface and an expedited data transfer capability is desirable. Group delivery should be selectable to support the binding process between a client and a server [53] since this can be more efficient than the use of global directory services. A reliable multicast service is not needed since most of the RPC systems do not guarantee such service. The quantitative service parameters, such as the requirements on throughput, delay, and bounded response time, depend on the specific application. With respect to the four service classes, the reliable real time class matches the basic requirements of RPC based applications [52]. If desired, the service can be enhanced with specific group delivery facilities.

4.3 Flexible Protocol Configuration

The provision of efficient protocol machines dynamically configurable from a pool of protocol building blocks and tailored to individual application service requests is considered as an important feature for forthcoming transport subsystems. Together with the described service interface, this provides a sound basis to serve the increasing variety of application requirements with adequate and efficient protocol machines.

According to the service specification, a *Protocol Configurator* selects the required protocol functions. Therefore, a protocol resource pool consisting of a set of protocol functions specified in a special description language, F-PDL [28], is accessed. The Protocol Configurator then generates a protocol machine description reflecting their dependencies. The instantiation of this description together with a protocol management agent is called a protocol machine. In case of pre-defined service classes, the protocol machine is pre-packaged and available to reduce the configuration delay. Subsequently, the protocol machine is mapped on the hardware platform. Dynamic changes of the protocol machine during an established communication relation may be needed, mainly due to changes in the application service requirements. A *Session Manager* handles requests for dynamic changes and decides about their eventual implementations. Moreover, the Session Manager takes care about inter-stream tasks, e.g., synchronization among streams. This is needed for applications that request multiple streams, such as data, video and voice streams. In this case, a separate protocol machine is configured for each stream and each direction of a stream.

The Protocol Configurator needs information about the availability of protocol resources as well as of local system and network resources required to implement a certain type of service. Several *Resource Management Tools* provide the required information possibly collected in collaboration with network management entities. The availability of system and network resources and their utilization are represented in corresponding data bases to which the Protocol Configurator has access.

4.4 Protocol Design

Within the presented framework, *Patroclos*, an inherently parallel protocol architecture is being designed [54], [55] using a function-based approach. The functional building blocks are specified as parallel finite state machines which cooperate via event signalling. The design of Patroclos is also heavily influenced by earlier experience gained with parallel software implementations of the XTP protocol on transputer networks [56]. Patroclos is specifically designed for an efficient implementation of an ATM transport subsystem on a hybrid multiprocessor architecture. Therefore, the communication overhead among the protocol functions is minimized to provide a high degree of concurrency. The Patroclos protocol architecture clearly separates send and receive functions to allow concurrent and independent processing. Furthermore, Patroclos distinguishes between transfer control functions and data transfer functions to decrease the influence of control processing on the regular data path. Transfer control functions do not operate on user data, they include, e.g., flow and congestion control, rate control, etc.). Data transfer functions comprise data manipulation functions (e.g., PDU composition), data analysis functions (e.g., duplicate control) and data control functions which decide about actions to be performed on the data units (e.g., resequencing). Patroclos could be used as a pre-defined protocol machine in the FCSS communication model. Flexible protocol configuration is supported by providing well defined interfaces among the different protocol functions which may form a part of the FCSS protocol resource pool.

5. CONCLUDING REMARKS

Emerging applications and forthcoming high performance cell-based networks may considerably change architecture, services and protocols of current transport systems. A brief survey of some current research work in this area has been given with a main focus on services and architectural issues. A variety of different design issues for an ATM transport subsystem are contained in the framework presenting a single integrated solution. The service interface and an efficient protocol architecture are highlighted in more detail.

To accommodate the performance requirements of emerging applications, efficient implementations of transport subsystems are required. For a prototype implementation of the presented framework, a hybrid multiprocessor platform is considered as an adequate implementation platform consisting of several general purpose processors as well as of dedicated hardware solutions for time critical functions, such as checksum calculation, timer management or buffer management. Parallel techniques implementing function level parallelism are applied [57]. A proper mapping of protocol machines on multiprocessor platforms is especially supported due to their fine granular and modular design around protocol functions specified as finite state machines. The first prototype is currently being implemented on a transputer based multiprocessor system.

Although this paper aims in presenting several increasingly important issues on future transport system designs, clearly, not all aspects are covered within the limited available space. For example, multicast which is a very important service feature in the future has not been discussed. However, modified protocol mechanisms and addressing schemes providing multicasting can easily be integrated in the presented framework. Generally, one may view future communication systems in terms of standardized configurable units consisting of protocol functions (e.g., comparable to the concept of Application Service Elements in higher OSI layers [58]). This would make them more open to changes in the networking environment than the current architectures with large-grained units formed of hierarchical layers. Furthermore, the important characteristics of flexibility and dynamics an be easily applied as shown by the presented framework.

REFERENCES

[1] W. Doeringer, D. Dykeman, M. Kaiserswerth, B. Meister, H. Rudin and R. Williamson, "A Survey of Light-Weight Transport Protocols for High-Speed Networks," *IEEE Transactions on Communications*, vol. 38, no. 11. November, 1990.

[2] D.R. Cheriton, "VMTP as the Transport Layer for High-Performance Distributed Systems," *IEEE Communications Magazine*, June 1989.

[3] D.C. Feldmeier, "An Overview of the TP+ + Transport Protocol Project," *in: A. Tantawy (ed), High Performance Communications*, Kluwer Academic Publisher, to appear 1993.

[4] W.T. Strayer, B.J. Dempsey, A.C. Weaver, *XTP - The Xpress Transfer Protocol*, Addison Wesley, 1992.

[5] M. Zitterbart, "High-Speed Transport Components," *IEEE Network*, pp.54-63, January 1991.

[6] M. Kaiserswerth, "A Parallel Implementation of the ISO 8802-2.2 Protocol," *IEEE TRICOMM '91, Chapel Hill, NC, USA*, April 1991.

[7] D. Clark, V. Jacobson, J. Romkey and H. Salwen, "An Analysis of TCP Processing Overhead," *IEEE Communications Magazine*, June 1989.

[8] T. Balraj, Y. Yemini, "Putting the Transport Layer on VLSI - the PROMPT protocol chip," *3rd IFIP Workshop on Protocols for High Speed Networks*, Stockholm, Sweden, May 1992.

[9] Esprit Project 5341, OSI 95 - High Performance OSI Protocols with Multimedia Support on HSLANs and B-ISDN - The Enhanced Connection-mode Transport Service of OSI 95, May 1992.

[10] ISO/IEC JTC1/SC6 N7309, Draft Guidelines for Enhanced Communication Functions and Facilities for the Lower Layers (For Comments), May 1992.

[11] IEEE, IEEE Std 802.6-1990, Distributed Queue Dual Bus (DQDB) - Subnetwork of a Metropolitan Area Network (MAN), 1991.

[12] Bellcore, Technical Advisory TR-TSV-000772, Issue 1, Generic System Requirements in Support of Switched Multi-megabit Data Service, May 1991.

[13] CCITT, Recommendations Drafted by Working Party XVIII/8 (General B-ISDN Aspects), June 1990.

[14] ATM Forum, ATM User-Network Interface Specification Version 2.0, June 1992.

[15] P. Mardsen, "Interworking IEEE 802/FDDI LANs Via the ISDN Frame Relay Bearer Service," *Proceedings of the IEEE*, vol. 79, no. 2, pp.223-229, February 1991.

[16] M. Kawarasaki, B. Jabbari, "B-ISDN Architecture and Protocol," *IEEE Journal on Selected Areas in Communications*, Vol. 9, No. 9, December 1991.

[17] W. Stallings, *Handbook of Computer Communications Standards, Vol. 2, Local Area Networks Standards, Second Edition*, Carmel: Howard W. Sams and Company, 1990.

[18] ANSI, X3T9.5, FDDI Token Ring Media Access Control, 1988.

[19] D. Comer, *Internetworking with TCP/IP*, New Jersey: Prentice Hall, 1991.

[20] International Organisation for Standardization, "OSI Basic Reference Model," *ISO 7498*, 1984.

[21] R. Haendel, M.N. Huber, *Integrated Broadband Networks - An Introduction to ATM-Based Networks*, Addison-Wesley, 1991.

[22] M. de Prycker, *Asynchronous Transfer Mode - Solution for Broadband ISDN*, Ellis Horwood, 1991.

[23] M. de Prycker, R. Peschi, T. van Landegem, "B-ISDN and the OSI Protocol Reference Model," *3rd IFIP WG 6.4 Conference on High Speed Networking*, Berlin, Germany, March 1991.

[24] W. Stallings, *ISDN and Broadband ISDN*, Macmillan Publishing Company, 1992.

[25] CCITT, Recommendations Drafted by Working Party XVIII/8 Recommendation I.363 B-ISDN ATM Adaption Layer (AAL) Specification, June 1992.

[26] CCITT, Recommendations Drafted by Working Party XVIII/8 Recommendation I.371 Traffic Control and Congestion Control in BISDN, June 1992.

[27] ISO, IS 8072, Information processing systems - Open Systems Interconnection - Transport Service Definition, 1986.

[28] M. Zitterbart, B. Stiller and A. Tantawy, "A Model for Flexible High Performance Communication Subsystems," *IBM Research Report, RC 17801*, February, 1992.

[29] A. Campbell, G. Coulson, F. Garcia, D. Hutchison, "A Continuous Media Transport and Orchestration Service," *ACM SIGCOM '92*, Baltimore, Maryland, August, 1992.

[30] B. Butscher, L. Henckel, T. Luckenbach, "The Communication Platform BERKOM for Multimedia Applications (in German)," *Informatik-Spektrum, Springer Verlag*, Vol. 14, 1991.

[31] B. Butscher, G. Goldacker, P. Todorova, "A Protocol Architecture for Transport Services in B-ISDN," *2nd IEEE Workshop on Future Trends on Distributed Computing Systems*, Cairo, Egypt, October 1990.

[32] W. Zimmer, "FINE: A High-Speed Transport Protocol Family and its Advanced Service Interface," *3rd IFIP Workshop on Protocols for High Speed Networks*, Stockholm, Sweden, May 1992.

[33] A. Danthine, "Esprit Project OSI 95 - New Transport Services for High-Speed Networking," *Computer Networks and ISDN Systems, Vol. 25*, November 1992.

[34] A. Danthine, Y. Baguette, G. Leduc, L. Leonard, "The Enhanced Connection-mode Transport Service of OSI 95," *4th IFIP Conference on High Performance Networking*, Liege, Belgium, December 1992.

[35] G. Blair, G. Coulson, F. Garcia, D. Hutchison, and D. Shepherd, "Towards New Transport Services to Support Distributed Multimedia Applications," *4th IEEE Workshop on Multimedia Communications*, Monterey, CA, April 1-4, 1992.

[36] D. Hehmann, M. Salmony and H.J. Stuettgen, "Transport Services for Multi-Media Applications on Broadband Networks," *Computer Communications*, 13(4), 1990.

[37] ISO, IEC 8073, Information processing systems - Open Systems Interconnection - Connection-oriented Transport Protocol Specification, 1988.

[38] ISO, IS 8073, Information processing systems - Data Communications - Protocol for Providing the Connectionless-mode Network Service, 1988.

[39] Z. Haas, "A Protocol Structure for High-Speed Communication over Broadband ISDN," *IEEE Network*, January, 1991.

[40] H.E. Meleis and A.N. Tantawy, "High Performance Networking and the Modular Communication Machine (MCM) Approach," *2nd IEEE Workshop on Future Trends on Distributed Computing Systems*, Cairo, Egypt, October 1990.

[41] N.C. Hutchinson, L.L. Peterson, "The x-Kernel: An Architecture for Implementing Network Protocols," *IEEE Transactions on Software Engineering, Vol. 17, No. 1*, January, 1991.

[42] S.W. O'Malley and L.L. Peterson, "A Highly Layered Architecture for High Speed Networks," *Protocols for High Speed Networks II, M. Johnson (ed.)*, Elsevier, 1991.

[43] D.C. Schmidt, D.F. Box, and T. Suda, "ADAPTIVE - A Flexible and Adaptive Transport System Architecture to Support Multimedia Applications on High Speed Networks," *Technical Report 92-46, University of Irvine, California*, 1992.

[44] Ch. Tschudin, "Flexible Protocol Stacks," *ACM SIGCOMM '91*, Zurich, 1991.

[45] E.W. Biersack, D.C. Feldmeier, "Transport Protocol Issues for ATM-based Networks," *EFOC/LAN, Munich, Germany*, June 1990.

[46] D.C. Feldmeier, E.W. Biersack, "Comparison of Error Control Protocols for High Bandwidth-Delay Product Networks," *2nd Workshop on Protocols for High Speed Networks*, San Francisco, CA, November 1990.

[47] E.W. Biersack, "Connection Management Using Synchronized Clocks," *3rd IFIP WG 6.4 Conference on High Speed Networking*, Berlin, Germany, March 1991.

[48] W. Bauerfeld, Horst. Westbrock, "Multimedia communication with high-speed protocols," *Computer Networks and ISDN Systems*, vol. 23, 1991.

[49] L. Svobodova, "Measured Performance of Transport Services in LANs," *Computer Networks and ISDN Systems*, vol. 18, 1989.

[50] A. Tang, S. Scoggins, *Open Networking with OSI*, Prentice Hall, 1992.
[51] O.G. Koufopavlou, A. Tantawy and M. Zitterbart, "Analysis of TCP/IP for High Performance Parallel Implementations," *17th IEEE Conference on Local Computer Networks*, Minneapolis, Minnesota, September 1992.
[52] M. Zitterbart, B. Stiller and A. Tantawy, "Application-Driven Flexible Protocol Configuration," *to appear in: KIVS '93, Munich, Germany*, March, 1993.
[53] S. Mullender (Ed.), *Distributed Systems*, ACM Press, 1989.
[54] T. Braun, "A Parallel High Performance Transport System for Metropolitan Area Networks," *5th IEEE Workshop on Metropolitan Area Networks*, Taormina, Italy, May 1992.
[55] T. Braun, M. Zitterbart, "Parallel Transport System Design," *4th IFIP Conference on High Performance Networking*, Liege, Belgium, December 1992.
[56] T. Braun, M. Zitterbart, "A Parallel Implementation of XTP on Transputers," *16th IEEE Conference on Local Computer Networks (LCN)*, Minnesota, Minneapolis, October 1991.
[57] M. Zitterbart, "Parallelism in Communication Subsystems," *in: A. Tantawy (ed.), High Performance Communications, Kluwer Academic Publishers*, to appear 1993.
[58] J. Henshall, S. Shaw, *OSI Explained*, Ellis Horwood, 1990.

XTP BUCKET ERROR CONTROL: ENHANCEMENT AND PERFORMANCE ANALYSIS*

Serge Fdida[1] and Harry Santoso[2]

[1] Laboratoire MASI, Université René Descartes-EHEI
45 rue des Saints Pères, 75006 Paris-France
Tel. 19-33-1-44-27-30-58, Fax. 19-33-1-44-27-62-86
email: fdida@masi.ibp.fr

[2] Laboratoire MASI, Université Pierre & Marie Curie
4 place Jussieu, 75005 Paris-France
Tel. 19-33-1-44-27-71-27, Fax. 19-33-1-44-27-62-86
email: santoso@masi.ibp.fr

ABSTRACT

Multicast is a major functional capability to be used in high-speed networks in order to support a wide range of applications. Transport layer multicasting is defined as a transfer of a Transport Protocol Data Unit from one sender to a well defined set of receivers. In this paper, we address the problem posed by the reliable service of the XTP multicast (we refer to such a service as statistical reliable) which is primarily related to error recovery issues. XTP error control scheme uses a bucket mechanism kept at the sender side. We discovered that the XTP multicast response to errors is inefficient. This drawback is caused by the multiple retransmissions of the same missing data packets. We show that the total number of redundant retransmissions may be proportional to the number of buckets. To overcome this problem, we introduce some other constraints in the sender side to prevent any unnecessary retransmission. However, some key parameters of the proposed algorithm have to be carefully sized for matching the application service requirements. We present a preliminary analysis of some parameters of the XTP enhanced bucket algorithm through simulation results. We mainly focus in the influence of the number of buckets in the system response time for two different network environments: FDDI and ATM based networks. We also address the problem of processing overheads at the sender side. We conclude that further work should be done in order to improve the efficiency of the algorithm for large multicast groups.

* This work is conducted within the framework of a joint CNET-CNRS project.

Asynchronous Transfer Mode Networks, Edited by Y. Viniotis
and R.O. Onvural, Plenum Press, New York, 1993

1. INTRODUCTION

Multicasting or peer-to-multipeer exchange is designed for transferring information from a single source to a well defined set of receivers. The multicast service is undoubtedly useful for distributed systems of the future. Multicasting offers significant performance advantages over multiple sequential unicastings in terms of efficient use of the network bandwidth, decreasing the elapsed time and increasing concurrency. It will thus rapidly become a preferred paradigm for future Group Inter-Process Communications [6] in which a single transmitting process sends a message to a group of well defined receiving processes. There are many systems that necessitate multicast services, they include: *resource finding* [3], *replicated file system* [23], *distributed games* [24], *distributed databases* [7], *email distribution list* [21], *routing table dissemination* [20] and *computer conferencing* [10], etc. In short, multicasting is extremely useful for many distributed applications.

The concept of multicast communications has received much recent attention. It might be considered as a complex and difficult problem. Most of the existing multicast algorithms are focused on the MAC or IP [25] layer. Little attention has been devoted to the development of multicast protocols for network or transport layers [11]. Motivated by the current research in this area and given the availability of a new communication technology based on Asynchronous Transfert Mode [13] (which inherently supports multicast and therefore necessarily needs to redesign the existing network and transport protocols), we thus focus in this paper on the design issues of the multicast transport protocol. More specifically, our study will be based on the XTP transfert[1] layer multicast. The reason for selecting this protocol as the basis of our study is that XTP is well documented [8,9] and employs many of the clever design concepts from the existing lightweight transport protocols [11,14,15,16]. See [18] for an excellent survey.

In a modern unicast transport protocol such as XTP [9] there are at least six procedures: connection management, data transfer, state/time synchronization, flow control, rate control, and error control. Incorporating such procedures in multicast transport layer imposes significant challenges and the new multicast protocols must be designed and integrated into the transport layer. One of the central issues associated with the development of the multicast transport protocol is error detection and recovery procedure. In this paper we concentrate on the issues of error control design for a lightweight transport layer multicast.

Note that XTP as well as VMTP are two examples of lightweight multicast transport protocols that allow destination users to receive in sequence one or more data-packets from one transmitter, but an absolute ordering of data-packets, sent by many transmitters, across all the multicast receivers is not guaranteed. Because of the inherent cost for ensuring absolute ordering [2, 4], it is thus in general unlikely to include this functionality within a lightweight transport protocol, it is left to the higher layer protocol to do so.

Multicast is a new functional capability for a high performance transport layer which is being studied by standard committees (ANSI, ISO, etc. [8, 12]). A transport layer multicast shall provide a simultaneous transfer of the same TPDU to a number of peer transport layer entities. In providing such a multicast service, the transport layer relies on the multicast capabilities provided by network layer. Each TPDU is designated as a multicast TPDU, it is then transfered to the network layer which provides for the generation and transmission of multiple copies of multicast TPDU (in the litterature, this capability is referred to as *packet replication*). Packet replication is extremely expensive and can seriously damage the overall lightweight transport protocol performance [5,17]. In order to define the logical relationship between the members of a group of Transport Service (=TS) users, the concept of *multicast conversation* is introduced, thus a transport multicast conversation is an association which may have more than two TS users. The conversation should be established by one TS user (called "initiator") with the associated peer entities before any multicast data transfer can occur. Within the transport layer, multicast conversation needs to support a wide variety of

[1] XTP transfer layer provides the facilities of the transport and network layers of the ISO Reference Model. The services supported by XTP are derived from the French GAM-T-103 transfer layer model for Military LAN.

TS user requirements. This requires that the *semantics* of transport layer multicast conversation be defined. We define three categories for this purpose.

Best Effort Multicast provides a "best effort" sequential transport of a single variable length TSDU to a set of TS users. The service is best effort in that missing or corrupted TSDUs are not delivered or recovered. All-reliable Multicast ensures that the lost or damaged TPDU is retransmitted until the transmitter knows with absolute certainty that all receivers have correctly received it. There exists another reliable semantic stronger than best-effort multicast but weaker than all-reliable multicast called statistical-reliable in which the reliability is enforced when desired by applications. The point at which the reliability becomes unacceptable would be application dependent. The Statistical-reliable Multicast does not enforce absolute reliability and the multicast sender does eliminate the need for maintaining state of individual multicast receivers. Using this semantic, the multicast transmission is considered successful if none, some or all TS users receive the multicast data [8,9]. As the transmitter is relieved of any concern with the states of receivers, there is no way to know whether or not all receivers have successfully received the multicast TPDUs. However, if a more stringent requirements on the reliability is needed, it can be implemented on top of this statistical-reliable multicast transport protocol, i.e. protocol that deals with reliability in an application dependent manner. The statistical-reliable should be efficient when the total number of destination population is large. One example of the transport protocol that provides a statistical-reliable multicast facility is XTP [9]. XTP multicast will be discussed in more detail in the following section.

The main contribution of this paper is to investigate and analyse the XTP bucket error control scheme. We point out the problem posed by the XTP bucket algorithm through an example and we also propose a solution to this problem so that the XTP multicast can work efficiently and eliminate the redundant retransmissions.

The paper is organized as follows. Section 2 presents the operation of XTP multicast, the bucket error control scheme for XTP, and an example illustrating its inefficiency. Section 3 gives the proposition to enhance the XTP bucket algorithm. Section 4 provides a preliminary performance analysis of the proposed algorithm for two network environments (FDDI and ATM). Section 5 concludes the paper.

2. THE XTP MULTICAST

The Xpress Transfer Protocol (XTP) was designed in order to merge the transport and network functionalities that can be easily implemented in hardware. The XTP multicast proposed two class of services, i.e. *acknowledged service* that ensures correct and sequential reception of packets and *unacknowledged service*. Fundamental to the development of the XTP multicast is the following assumptions: 1) the underlying layer(s) must support the basic multicast service, 2) the multicast transmitter must not have any knowledge of the explicit address of the receiver set, thereby eliminating the need for maintaining state of individual multicast receivers, 3) the receivers are not allowed to send any data-bearing packet back to the sender, thus the data transfer is only simplex and 4) the multicast transmitter controls all state information exchanges between transmitter and receiver set, and the receivers should not generate any control message without explicit command from the transmitter. The specification of XTP explicitly describes how to provide flow-, rate-, error- control mechanisms in the transport layer multicast and the proposed heuristic known as *"bucket algorithm"* allows XTP to efficiently support such mechanisms.

2.1. The Operation of XTP Multicast

The XTP multicast can accomodate two different types of reliable semantics, i.e. statistical reliable and best-effort semantics. The interpretation of statistical reliable here is that lost or erroneous XTP packets will be retransmitted. It is said that the multicast transfer is error controlled. With best effort semantic, the receivers make no effort to recover lost or damaged XTP packets. The multicast sender may choose to turn off the retansmission

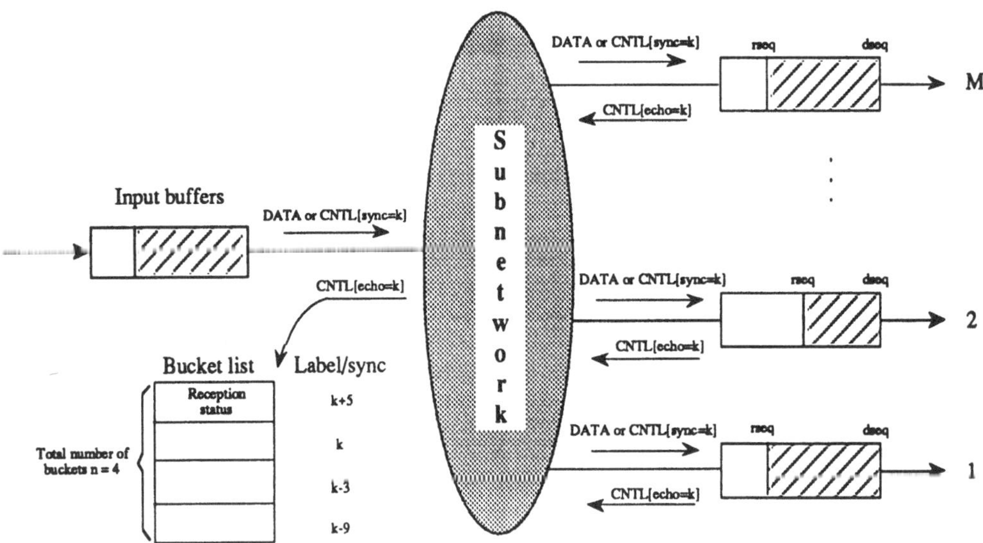

Figure 1: Multicast Operation in XTP

procedure by setting the NOERR bit in the header field of the packet. In this paper, we are only concerned with the statistical reliable multicast and more specifically we will address the issues that arise in XTP multicast error control.

Figure 1 shows that a sender multicasts the data-bearing and control packets (in XTP parlance, DATA or CNTL packets) to a set of M receivers using a *single group address*. The receivers are not allowed to send DATA packets in the return direction. XTP assumes that the multicast sender has neither knowledge of all the receivers nor the individual addresses of all receivers. Also note that XTP multicasting requires that the underlying layers, e.g. IEEE-LAN MAC, FDDI MAC, or IP layers, provide basic multicast services but not necessarily use a broadcast media. This underlying layer is envisaged to be used primarily for packet transport and therefore the XTP packet will be encapsulated within either MAC frame or IP datagram.

The XTP's multicast error control is based upon the concept that the sender periodically multicasts a CNTL[*sync=k*] packet to the receiver set to which each receiver immediately responds by multicasting a CNTL[*echo=k*] packet to all member of the group. Each receiver must copy the *sync* value from incoming CNTL[*sync=k*] packet into the *echo* field of outgoing CNTL[*echo=k*] packet returned to the sender. *Sync-echo* pair is used to correlate a request CNTL[*sync=k*] and its associated response CNTL[*echo=k*], this procedure is known as *synchronizing handshake*. This will keep sender and receivers synchronized with respect to state information. The CNTL[*sync*] is used to solicit the status of the receivers and CNTL[*echo*] is used to transfer "status information" about the receiver. Each "status information" may contain *reception status, flow control, rate control, and roundtrip time estimate, etc.* In this paper, we focus only on the reception status which is represented by two variables *rseq* (received sequence number) and *dseq* (delivered sequence number). The *dseq* and *rseq* represent two kinds of acknowledgement in XTP. The *dseq* value is used by the receiver to indicate one larger than the sequence number of the received DATA packet last delivered to the XTP user. The *dseq* value is the only information required to release the sender's input buffers. The *rseq* contains a value one larger than the largest monotonic sequence number of the received DATA packet received without error. The *rseq* provides the retransmission information to the sender which is necessary to trigger a Go-back-n retransmission, starting with the byte identified by *rseq*.

Since the sender is not aware of the individual addresses of the receiver set, it can not know from which receivers a burst of CNTL[*echo*] packet is received. A mechanism is required to efficiently interpret this burst of CNTL[*echo*] packets and to determine the time period that the sender should wait for collecting CNTL[*echo*] packets before updating its state. To accomplish these objectives, XTP specifies a heuristic[2] known as the ***Bucket Algorithm*** kept at the sender. The buckets can effectively accumulate the CNTL[*echo*] packets coming from multiple receivers and then forward the summary of CNTL[*echo*] packets to the sender's state machine. In other word, a multicast sender that executes a bucket algorithm treats all the receivers as a single entity thus many receiver's retransmission requests are transformed into a single retransmission. We believe that for a large receiver set, the bucket algorithm appears as a good solution. We address the bucket mechanism in more detail in the next subsection.

2.2. Bucket Error Control for XTP

In this subsection, we give a specification of XTP's bucket error control mechanism which is depicted in figure 2. XTP multicast provides the procedure that allows the multicast sender to capture the snapshot of the receivers' status on a periodic basis. This states capturing procedure begins when the sender generates CNTL[*sync*] packet and the receivers respond with CNTL[*echo*] packets[3]. Once a multicast transfer is initiated, a sender is imposed to maintain *a bucket list* consisting of *a series of buckets* which correspond to a finite time period T_p (T_p might be assigned with the worst case estimate of network-roundtrip-time or greater as suggested in [9]). *A bucket* may correspond to a time window T_w where its length will determine the interval between the generation of two consecutive CNTL[*sync*] packets initiated by the sender. In figure 2, the bucket *k* is created when the sender submits a CNTL[*sync=k*] packet. The bucket *k* will be used to store all response CNTL[*echo=k*] packets associated with CNTL[*sync=k*]. For a given sender, the total number of buckets *n* should be (T_p/T_w) buckets. We note that T_p and T_w are two design parameters to be tuned.

The parameter T_p bounds the time taken to recover from errors in one-to-many multicast communication, it can also be defined as the time interval over which the multicast sender anticipates receiving all receiver responses. Obviously, the larger the value of T_p the higher the reliability, giving the receivers a higher opportunity to recover from the missing data. Unfortunately, the higher value of T_p has a negative impact on the throughput of the multicast communication, this is because the sender's input buffers cannot be released more quickly. For this reason, the value of parameter T_p should be tuned relative to the desired reliability and the throughput performance. A sender should be capable to decide its own tradeoff between reliablity and throughput condition. As described previously, the bucket algorithm also uses another time window T_w and its value is typically shorter than T_p. The T_w is the time interval between two control packet transmissions and therefore the choice of this interval is an important factor, affecting the freshness of status information content received from receiver set. The smaller value of T_w guarantees that the information will be fresh and that the protocol will be responsive. But it will also result in a larger bandwidth overhead, higher sender's and receiver's nodes processing costs. Again, a multicast sender should be able to choose the cut off between protocol responsiveness and processing/bandwidth overheads.

As we have seen, for each submission of CNTL[*sync*] packet a new bucket is created, this newly created bucket is labeled using the value of *sync* sent out in CNTL[*sync*] packet. Therefore, a range of *sync* values sent out in CNTL[*sync*] packets will identify a series of buckets. At the sender side, the *echo* value of the incoming CNTL[*echo*] packet will be associated with the bucket label and the most conservative values of status information carried in the stream of CNTL[*echo=k*] packets will be stored in a bucket whose label is *k*. Note that any received CNTL[*echo*] packet with an *echo* value that cannot be associated with the bucket

---

[2] In fact, there are two other heuristics called *Damping* and *Slotting*. *Damping* is implemented by receivers to prevent the redundant response CNTL[*echo*] packets from being transmitted. *Slotting* is a technique to schedule the response CNTL[*echo*] packet at some random time by the receivers. The goal is to avoid the arrival of back-to-back CNTL[*echo*] packets from many receivers that may cause congestion at the multicast sender. See [9] for the detail.

[3] It should be noted that the receivers may not generate any information back to the sender without specifically requested by the multicast sender.

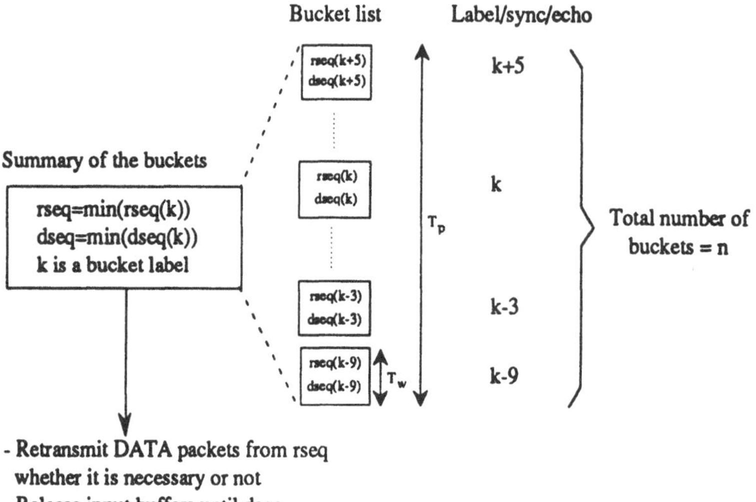

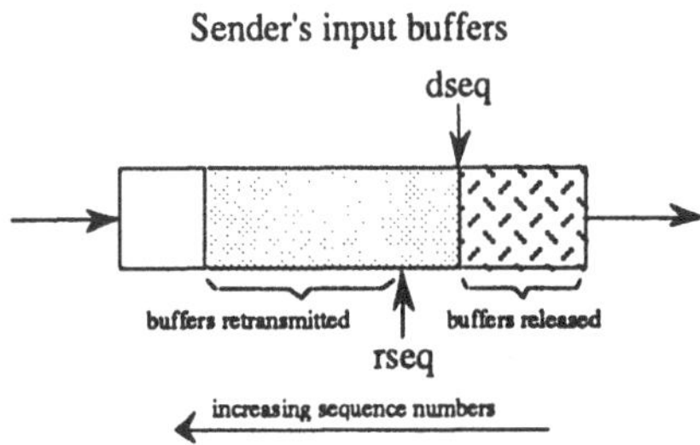

Figure 2: Bucket Error Control in XTP

will be ignored and this will not violate the protocol. As an example, suppose that the sender issues a CNTL[*sync=3*] then the bucket with label=sync=3 is created right away. The bucket 3 will consist of the *minimum* of all values for *dseq* and *rseq* carried in all response CNTL[*echo=3*] packets.

Of course, in order to uniquely identify the buckets, the *sync* value must be incremented by one for each (re)transmission of DATA or CNTL[*sync*] packet. This *differs* from the XTP specification in that the *sync* is only incremented for every transmitted DATA packet (see [9] pages 22 and 80) and not for CNTL[*sync*] packet. However, since the CNTL[*sync*] packets shall be transmitted even if no DATA packets are (re)transmitted, under such a scenario no new bucket can be created if the *sync* value is not also updated when CNTL[*sync*] packets are sent. Therefore, we also need to update *sync* value each time the DATA packet is submitted.

At the moment when a new bucket is created and if all the bucket list is *full*, the oldest bucket (i.e. the one labeled with the lowest value of *sync*) must be released in order to create a new one. Note that at this point in time, the sender's state should also be updated by applying the *minimum* values of *dseq* and *rseq* over *all the buckets* in the bucket list. Then the sender begins retransmitting with the byte identified by *rseq* and releases all input buffers containing data whose sequence numbers are less than the *dseq* value (see lower part of figure 2). The same action will take place each time the bucket is recycled.

2.3. Example & Analysis

For the purpose of providing a concrete example of bucket mechanism, a simple DATA and CNTL packets exchange diagram is shown in figure 3. It includes one sender and two receivers. The sender executes the bucket algorithm using two buckets. Sender's input buffers, bucket contents and progress of multicast transfer are listed along the vertical line at the sender side. The sender's input buffers hold the DATA packets for possible (re)transmission at some future time. Any horizontal dashed line indicates a time boundary at which the multicast sender is transmitting the CNTL[*sync*] packet, creating or recycling the bucket, and applying the bucket variables to update the sender's state (i.e. release the input buffers and retransmit the missing data). The distance between two horizontal dashed lines represents the period between the two control packet transmissions or the so-called time window T_w.

Consider now the scenario presented in figure 3 in which the multicast sender transmits four DATA packets to a set of receivers {1,2}. The first containing DATA0=[0...99], the second containing DATA1=[100...199], the third containing DATA2=[200...299], and the fourth containing DATA3=[300...399]. The transmission starts with the value of *sync=1*, this *sync* value is incremented for every (re)transmission of DATA or CNTL packet. In this study, we assume that the DATA[*sync=2*] destined for receiver 2 and the CNTL[*echo=3*] from receiver 2 get lost. We use this scenario to describe how retransmission is handled in XTP multicast implementing bucket algorithm.

During the first two time windows T_w (see figure 3), the buckets labeled with 3 and 6 are created. The reception status (*dseq* and *rseq*) contained in response CNTL[*echo=3*] packets will be accumulated in bucket 3 and the sender will only record the minimum *dseq* and *rseq* values over all incoming CNTL[*echo=3*] packets. The same rule will be applied for every other buckets.

When the sender generates CNTL[*sync=7*] packet at the beginning of the third time window, the bucket list is full (i.e. occupied by buckets labeled with 3 and 6) then the sender's state must be updated by taking the minimum *rseq* and *dseq* selected from the buckets 3 and 6. Right away, the oldest bucket (label=3) must also be emptied and replaced by the newest bucket (label=7). In the example, the selected minimum *rseq* is 100 (it is taken from bucket 6) and the selected minimum *dseq* is 100 (by chance, it is also taken from bucket 6). The minimum *rseq*=100 indicates that the sender should retransmit all DATA1, DATA2 and DATA3 packets with sequence numbers from 100 to 399. The first retransmission of these three DATA packets occurs during the third time window. The minimum *dseq*=100 serves to release the input buffers for DATA0 packet with sequence number less than 100. Observe that the bucket 6 containing the selected minimum *rseq*=100&*dseq*=100 values remain in the bucket list.

At the start of the fourth time window, as before, the sender's state will be updated and the bucket 6 is replaced with the bucket 11. As per the protocol rule, on updating its state the sender uses the minimum values *rseq*=100 & *dseq*=100 as they are still the worst case values. In fact, these values are out-of-date and have been used to conduct the first retransmission. Therefore, it will result the second retransmission which is in fact redundant.

When the fifth time window starts, again, the third redundant retransmission cannot be avoided. As we can see from the diagram in figure 3, at the time the sender decides to perform the third retransmission, it uses the worst case estimate of state information (i.e. CNTL[*echo=7*] with *rseq*=100 and *dseq*=100) sent out by receivers 2 before the first retransmission occurs. Of course, this state information is no longer valid. XTP bucket error control mechanism cannot preclude this from happening.

The scenario depicted in figure 3 illustrates that the XTP bucket error control is not efficient due to the repeated unnecessary retransmissions. In the extreme case, the number of redundant retransmissions is as many as the total number of buckets. Thus the retransmission will be performed once per bucket until the worst case values, being used to make the loss decision, are gone. One solution to this problem will be presented in the next section.

3. ENHANCED BUCKET ERROR FOR XTP

One disadvantage of XTP bucket mechanism is that the missed DATA packets may be retransmitted multiple times. The number of these redundant retransmissions may be proportional to the total number of buckets as pointed out earlier. It can be seen that the cost of using this mechanism is loss of bandwidth due to unnecessary retransmission of DATA packets. The cause of this repetitive retransmissions can be explained as follows.

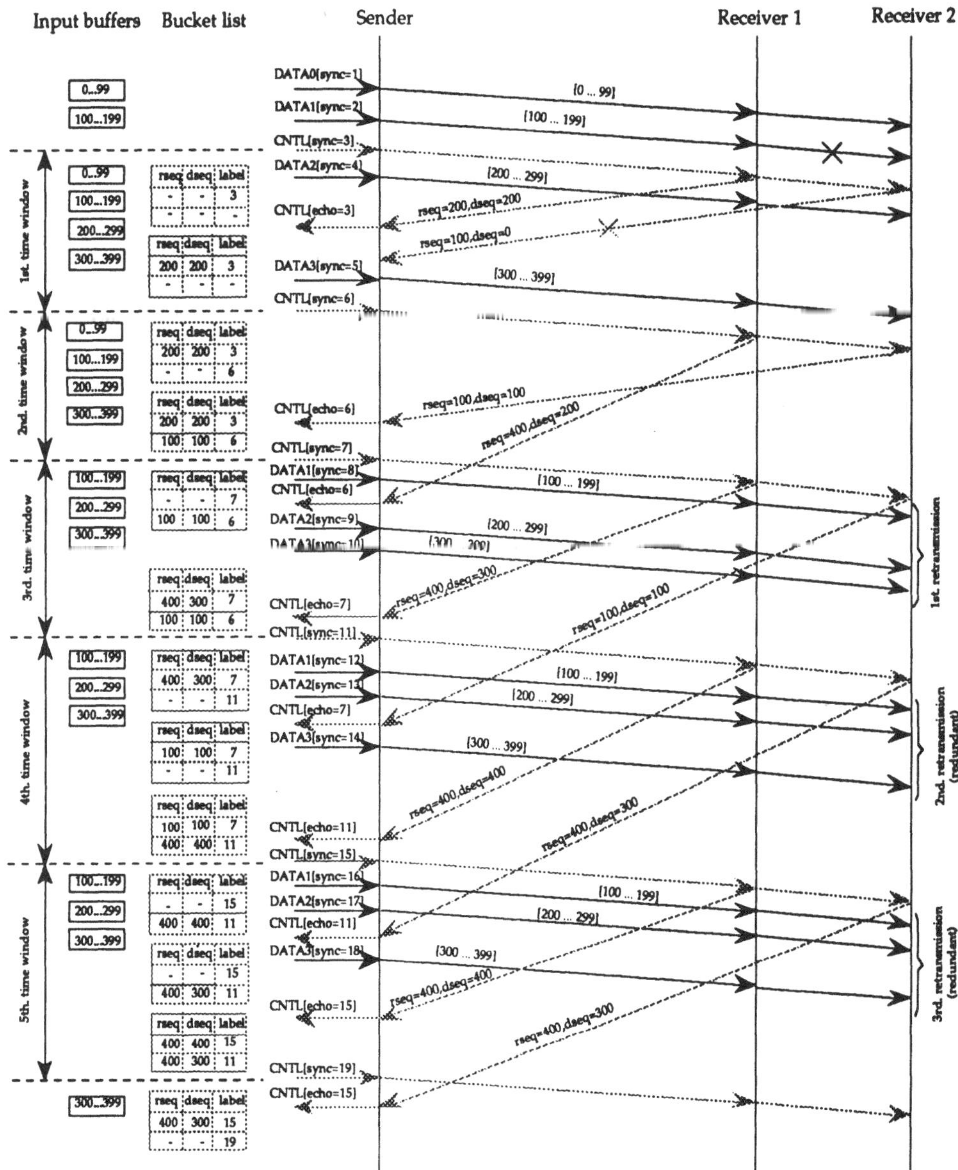

Figure 3: Scenario Illustrating the Redundant Retransmissions in XTP Bucket Algorithm

Whenever making a loss decision, the XTP multicast sender relies on the worst case estimate of the state information selected from a bucket list. The sender is unaware if the state information actually applied to make the loss decision is the one that has been used in the previous time windows.Consequently, the sender may retransmit multiple times the same missing data, leading to inefficiency. This inefficiency could be avoided by ensuring that the sender has the capability to recognize the worst case estimate of the most up-to-date state information. In this way, the sender does not retransmit the missing data packets until it is sure that this action will not cause redundant retransmission of the same missing data. It is not difficult to include this capability in the XTP bucket algorithm as will be developed in this section. We will refer the bucket algorithm with this capability as the *enhanced bucket algorithm* for XTP.

Another problem is that the XTP bucket algorithm does not explicitly discuss how the multicast sender can know whether there are any data gaps at the receiver set. Using the *rseq* alone for that purpose is not sufficient. Some additional information may be required by the multicast sender. For this reason, we introduce another bucket variable *lseq* which contains a value one larger than the largest sequence number of DATA packet ever (re)transmitted as known at the beginning of the bucket creation. In reality, this local information is readily available at the sender side. By comparing the values of *rseq* and *lseq*, the multicast sender will be able to tell if there is any missing data at some receiver(s). For the completeness of the algorithm, we also introduce the checking procedure to determine whether the requested data for retransmission still resides in the input buffers. This is based on the comparison between the saved worst case value of *dseq* (i.e. *Saved_min_dseq*) and the current worst case value of *rseq* (i.e. *Min_rseq*). The missing DATA packet can still be recovered if *Min_rseq* ≥ *Saved_min_dseq*.

In specifying our enhanced bucket algorithm, we make use the following variables maintained at the sender side as illustrated in figure 4:

1. A bucket is said to be in error if the *lseq* value does not match with *rseq* value (*lseq*≠*rseq*). All labels of the buckets in error are listed in array *L*. Thus, *L* is an array containing the bucket's label(s) and it is empty if there is no bucket in error in the bucket list.
2. *Min_dseq* is a variable containing the minimum value of *dseq* selected from all the buckets in the bucket list.
3. *Min_rseq* is a variable containing the minimum value of *rseq* selected from the buckets whose labels are in the array *L*. The label of the selected bucket is saved in another variable *k_error*. Therefore, *k_error* is a member of *L*.
4. *Saved_min_dseq* (*Saved_min_rseq*) is a variable used to store the current value of *Min_dseq* (*Min_rseq*).
5. Recall that the *Sync* value is incremented by one for every transmission of DATA or CNTL packet and every retransmission of DATA packet.

So far, we have presented a method of collecting control packets containing receivers' states using bucket mechanism in XTP multicast. In that, the state information being used to perform retransmission decision is taken from a bucket selected in the bucket list. We discovered that this retransmission procedure could not prevent multiple retransmissions of the same missing packets. In order to overcome this inefficiency, at time-window T_w expiration, we must apply the following retransmission rules:

(1) Examine each bucket to see if *lseq*≠*rseq*. Once the sender has found the buckets for values *lseq*≠*rseq*, it places the label(s) of the(se) bucket(s) in an array *L*. The array *L* is empty if no bucket holds *lseq*≠*rseq*.
This rule ensures that the retransmission is carried out if some receiver(s) missed their DATA packets.

(2) Select one bucket label among those listed in array L. The selected label must correspond to the bucket which holds the minimum value of *rseq*. The sender puts the selected label and the minimum value of *rseq* in another variables *k_error* and *Min_rseq* respectively.
Note that after each retransmission, the *sync* and *Min_rseq* values are also saved in variables *Saved_sync* and *Saved_min_rseq* respectively. The *Saved_sync* and *Saved_min_rseq* are used to remember the last *sync* value and the starting sequence number of the previously retransmitted DATA packets.

(3) The retransmission is performed only if the array L is not empty, the requested data is still retained in the input buffers (*Min_rseq* ≥ *Saved_min_dseq*) and one of the following conditions is satisfied:
- The *Min_rseq* < *Saved_min_rseq*
This condition implies that the starting sequence number of the currently requested retransmission is less than the one contained in the previously retransmitted DATA packets.
- The *Min_rseq* ≥ *Saved_min_rseq* and *k_error* > *Saved_sync*
The condition *Min_rseq* ≥ *Saved_min_rseq* serves to check if the starting sequence number of the currently requested retransmission is greater than or equal to the one contained in the previously retransmitted DATA packets. If so, in order to prevent unnecessary retransmissions the constraint *k_error* > *Saved_sync* should also be satisfied.

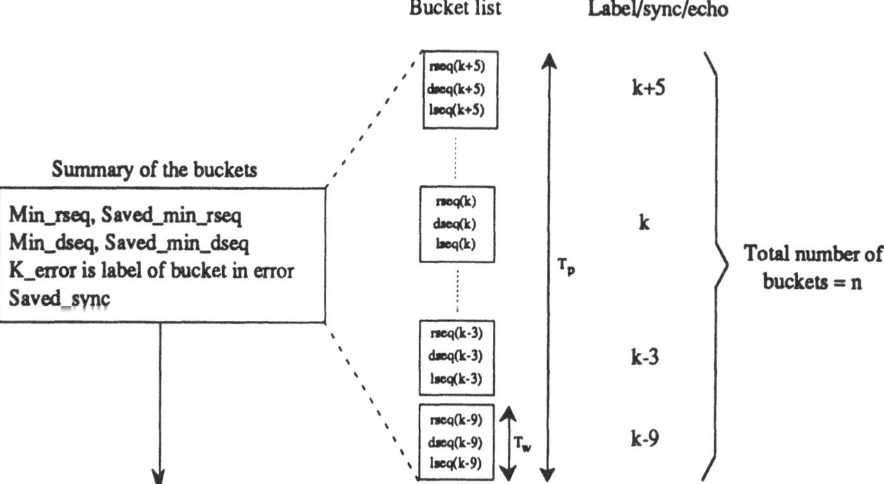

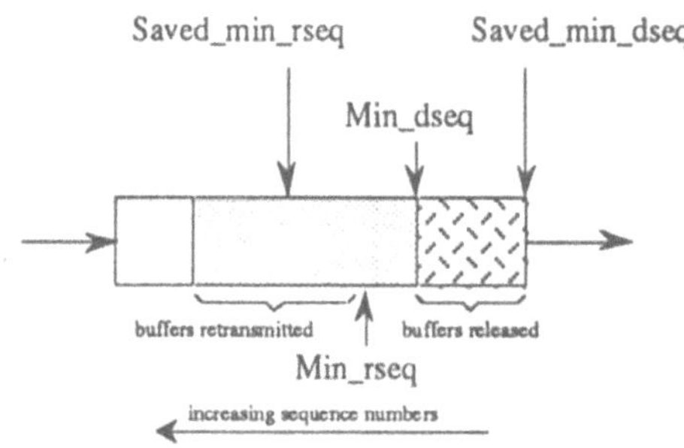

Figure 4: Enhanced Bucket Error Control for XTP

The condition *k_error > Saved_sync* indicates that the retransmission will be reattempted if the response control packet(s), associated with request control packet(s) submitted by the sender after the retransmission, is (are) received by the sender. Thus, the sender will be sure that the state information taken from the bucket for updating its state is most up-to-date.It is important to note that for the purpose of the correctness of the proposed algorithm, *Saved_min_rseq* must be initialized with the largest sequence number of the DATA packet ever transmitted just before the sender's state is updated for the first time and *Saved_min_dseq* and *Saved_sync* must be initialized with zero.

(4) If Min_dseq > Saved_min-dseq then the input buffers for data with sequence numbers less than Min_dseq may be released.

The complete procedures of the enhanced bucket algorithm will be listed in the sequel. Most of the procedures is self explanatory.

Let us now apply the enhanced bucket algorithm to the scenario presented in section 3, the resulting time diagram is shown in figure 6:

At the beginning of the third time window, the sender observes the following: L={6}, k_error=6, Min_rseq=100, Min_dseq=100, Saved_min_rseq=400, Saved_min_dseq=0, Saved_sync=0. As these values satisfy the retransmission conditions, the first retransmission takes place. At the end of this time window: Saved_min_rseq=100, Saved_min_dseq=100, Saved_sync=10.

At the start of the fourth time window, the sender uses the following state variables: L={6}, k_error=6, Min_rseq=100, Min_dseq=100, Saved_min_rseq=100, Saved_min_dseq=100, Saved_sync=10. Because these values do not satisfy the retransmission conditions (here k_error < Saved_sync), the sender will do nothing at this time window. It is easy to see that this is an improvement over XTP bucket algorithm, the unnecessary retransmission is avoided.

The complete procedures of the enhanced bucket algorithm are described in figure 5.

```
If array L is empty then do Input Buffers Release procedure
else do Retransmission procedure


Retransmission procedure:
  If (Min_rseq ≥ Saved_min_dseq) then
  & recovery is possible: data still resides in the input buffers &
        Begin
           If (Min_rseq < Saved_min_rseq) then
           & retransmission is necessary &
                     Begin
                         Input Buffers Release procedure;
                         Saved_min_rseq := Min_rseq;
                         Retransmission from Min_rseq then Saved_sync := sync;
                     End;

           Else  If (Min_rseq ≥ Saved_min_rseq) and (k_error > Saved_sync) then
                 & retransmission is not redundant if conditions are satisfied &
                     Begin
                         Input Buffers Release procedure;
                         Saved_min_rseq := Min_rseq;
                         Retransmission from Min_rseq then Saved_sync := sync;
                     End;
                 Else    Input Buffers Release procedure;
        End;

  Else  If (Min_rseq < Saved_min_dseq) then
        & recovery is not possible &
                   Send notification message informing the receivers that the message
                   with sequence number identified by Min_rseq cannot be retransmitted,
                   the receiver(s) should leave the multicast conversation or take
                   any other action.


Input Buffers Release procedure:
  If Min_dseq > Saved_min_dseq then
  & release input buffers &
                     Begin
                         Saved_min_dseq := Min_dseq;
                         Release the input buffers using Saved_min_dseq;
                     End;
```

Note:

- Saved_min_rseq must be initialized with the largest sequence number of the DATA packet ever transmitted just before the sender's state is updated for the first time.
- Saved_min_dseq and Saved_sync must be initialized with zero.

Figure 5: Procedure of the enhanced bucket algorithm

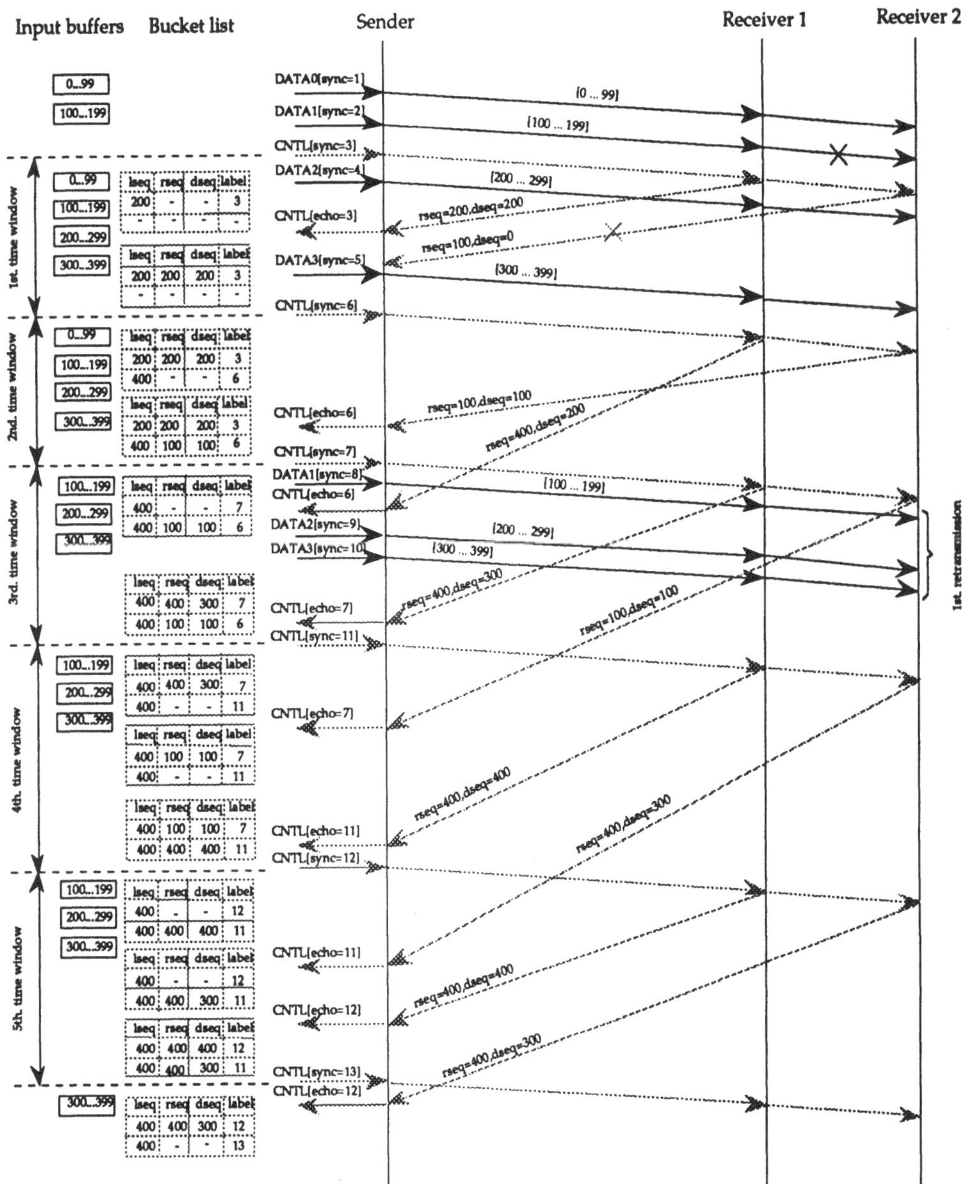

Figure 6: Scenario Illustrating the Efficiency of Enhanced Bucket Algorithm

4. PERFORMANCE ANALYSIS

This section briefly discusses our preliminary results on the simulation analysis of XTP enhanced bucket algorithm. As described in section 2.2, the XTP multicasting is defined as being statistical reliable. In order to guarantee an acceptable level of reliability, a carefull setting of some protocol parameters is required. Also, the protocol responsiveness and conserving processing should be another objectives in choosing these parameters.

There are 1 multicast sender and N receivers in a multicast group. Each sender or receiver is under one CPU control and each has a buffer capacity of B bytes. In this analysis, we assume that the underlying networks are FDDI and ATM, characterized by a negligible low error rate. Thus, the majority of error is caused by the buffer overflow. For FDDI network, we consider a ring of length L with data rate of 100 Mbit/s. The maximum frame size of FDDI is 4500 bytes. Each receiver is located equi-distantly on the ring such that the *i-th* receiver is on the location [i/(N+1)]L from the sender station. For ATM network, the distance between sender and *i-th* receiver is determined by end-to-end delay of R_i, where R_i is supposed to be uniformly distributed in the interval [R_{min}, R_{max}]. The transfer rate of ATM network is 155 Mbit/s.

The sender generates DATA traffic. The interval between consecutive DATA packets is exponentially distributed with mean value S_c. The size of generated DATA packets follows a uniform distribution with mean [500,4500] bytes. The receivers receive DATA packets multicast to them by the sender. In order to reflect the changes in loading conditions at the receivers, we explicitly model these variations by introducing the background traffic to each receiver. The inter-arrival time of the background traffic has also an exponential distribution with mean S_b. The mean packet size of background traffic is uniformly distributed within [500,4500] bytes.

In performing the evaluation on the bucket mechanism, one must take into account the effect of packet processing overheads, and they are defined as follows:

- Processing time of CNTL packet = X
- Processing time of DATA packet = 3X
- Processing time of ECHO packet = Y

Our simulation experiments focus mainly on the response time for data packets. The response time is the mean elapsed time between the submission of a DATA packet at the sender and its forwarding to the users at all the receivers. In this study, we will only investigate the interesting interactions between the response time (RT), the number of receivers (N), the round-trip time (RTT) and the number of buckets (NBUCKETS).

The following parameter values are used in the generation of figure 7: FDDI network with L=100 Km and propagation delay of 5 μs/km, RTT=7 ms, B=15000 bytes, X=0.5 ms, S_c=15 ms, S_b=7 ms, and NBUCKETS=1. Figure 7 compares the response time performance as a function of N for different values of Y (i.e. Y=0.1; 0.3; 0.5 ms). Figure 7 indicates that all three curves are overlapped when the number of receivers $N \leq 5$. As N increases, the three response time curves diverge, because the sender CPU is highly utilized to process more ECHO packets. As ECHO packet has a higher priority than DATA packet, the sender has a less time to send off the backlog DATA packets into the network. Roughly speaking, the sender CPU has to process (N*NBUCKETS) number of ECHO packets for each time window T_w. It is apparent that the parameters N and NBUCKETS limit the performance. In practice, there are two possible solutions for this problem. The first is to decrease the number of buckets, however, this solution is not always possible due to the reliability requirement. The second solution is to use the slotting and damping procedures. Using these mechanisms, the receiver is prevented from sending a redundant ECHO packet to the multicast group.

The parameter values used to plot figure 8 are the same as in figure 7 with the following exceptions: RTT[4] is varied (i.e. RTT=6; 7; 8; 9; 10 ms), these different values of RTT would model the situation where some receivers change their status due to the loading conditions. Note that for a fixed NBUCKETS, increasing RTT is equivalent to increasing T_w (i.e. interval between two CNTL packets). We choose Y=0.3 ms, NBUCKETS=1. Figure 8 shows the effect that round-trip time RTT has on the response time performance. Number of receivers is plotted on the horizontal axis versus response time on the vertical axis. From the graph it is clear that the larger the value of RTT, the smaller the response time. This phenomenon can be explained as follows. For a fixed number of buckets, when RTT increases the sender CPU has a more available time to process the backlog DATA packets

[4] RTT is defined to be that required for the CNTL packet to traverse the network, to be received at the receiver protocol entity, to traverse back the network, and then to be received again by the sender.

then submit them to the network. This would be, of course, decreasing the DATA packet waiting time at the sender side. However, it is also very important to note, for a fixed number of buckets the larger value of RTT has a negative impact on the DATA packet loss at the sender buffer (it is not shown in the figure 8). This is because the sender buffer is released more slowly when the RTT increases.

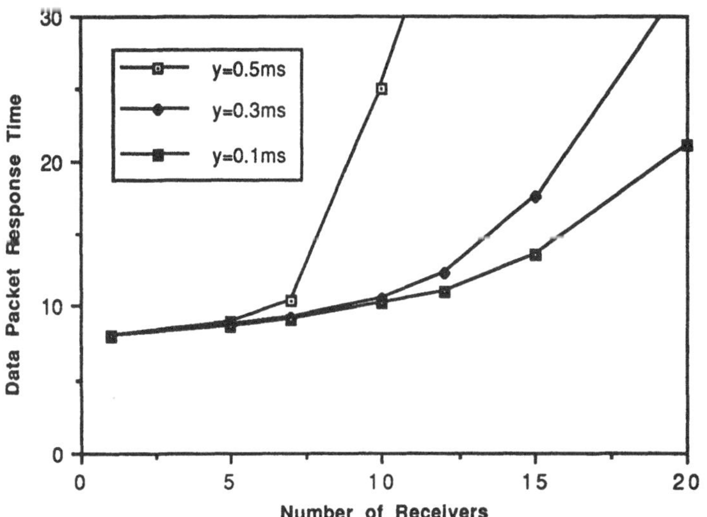

Figure 7: The Response Time Performance versus Number of Receivers for FDDI (with Y-parametrized plots)

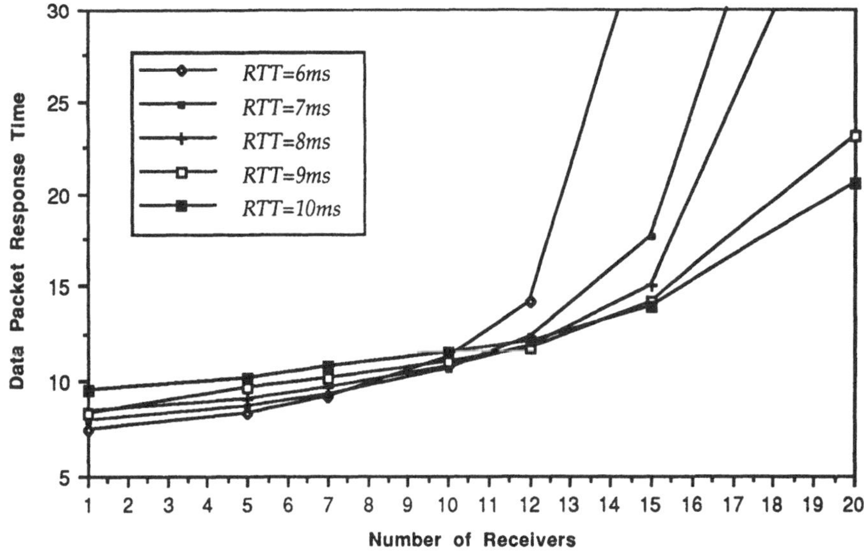

Figure 8: The Response Time Performance versus Number of Receivers for FDDI (with RTT-parametrized plots)

In order to see how the response time performance depends on the number of buckets NBUCKETS, we tested our simulator on the ATM environment. The network under study is defined by a round-trip time uniformly distributed in the interval [15ms, 28ms]. In our simulations, T_p is set to 50ms. In this situation, we have large sender and receivers buffer size. The data packet response time is sentitive to the number of buckets. We observe (figure 9) that there exist a minimum value around 15 buckets in our case. The ground traffic increases the data response time and loss at the output as well. We also observe that the number of input and output loss increases when the background traffic is too high and the number of buckets too small. That is increasing the number of buckets can decrease the losses at the stations. Also, we have focused on the reliability using a parameter named the Non Recoverable Errors (NRE). NRE is used to measure the reliablity and is defined as how many times the sender can not recover the packet loss requested by receiver set. We have seen that NRE is negligible for the two environments under study.

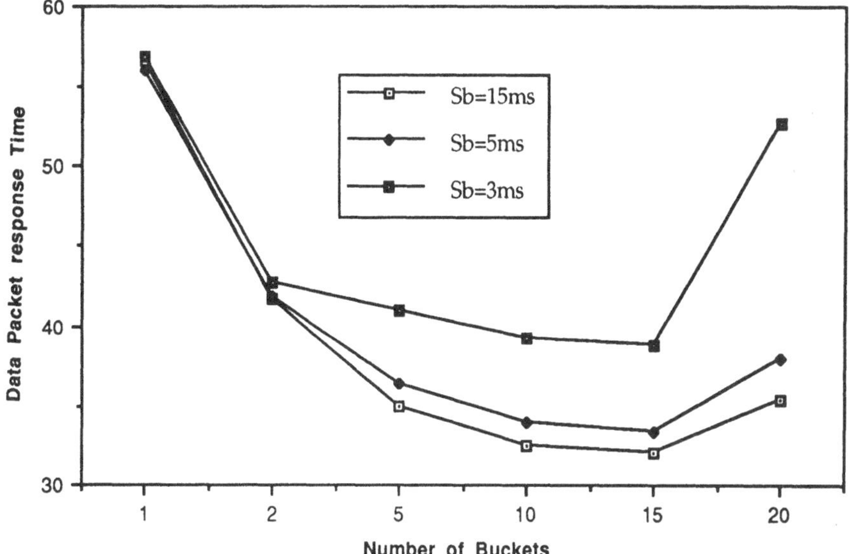

Figure 9: Influence of the number of buckets on the data packet response time (ATM Network) Sb= background traffic to every receiver, Sc= 15ms, Buffer size= 100000bytes, y=0.1ms

5. CONCLUSION

We have described the operation and characteristic of the XTP's multicast error control system using bucket mechanism. This class of error control is functionally simple and well suited for controlling error in a large receiver set. We refer to such a semantic that can be supported by XTP bucket error control algorithm as statistical reliable semantic. The statistical reliable semantics strongly depends on the two parameters T_p and T_w. For a given reliability requirement, the application should appropriately set these two parameters though this process is not easy.

We have also presented the inefficiency of XTP bucket algorithm when responding to multicast errors. This may result in a possible loss of bandwidth due to the redundant retransmissions. We proposed a modification to overcome this inefficiency by introducing another constraints to avoid any unnecessary retransmission. Moreover, this technique can be efficiently used as a uniform unicast and multicast communication, as opposed to the current XTP.

We have presented preliminary results for sizing the two parameters T_p and T_w of the XTP bucket error algorithm. We mainly focus on the influence of the number of buckets in the system response time for two different network environments: FDDI and ATM based networks, and also adress the problem of processing overheads at the sender side. These preliminary results show that further work should be done in order to improve the efficiency of the proposed algorithm for large multicast groups.

Acknowledgement

We would like to thank Maria Kihl and Katti Sundelin from Lund Institute of Technology (Lund, Sweden) for programming and running the simulator.

6. REFERENCES

[1] ISO-8072, Information Processing Systems, Open Systems Interconnection, Transport Service Definition, 1988.

[2] J. Chang and N.F. Maxemchuk, "Reliable Broadcast Protocols", ACM Transactions on Computer Systems, vol. 2, no. 3, pp. 251-273, August 1984.

[3] Mustaque Ahamad et. al., "Using Multicast Communication to Locate Resources in a LAN-Based Distributed System", Proc. of IEEE 13th. Conference on Local Computer Networks, pp. 193-202, October 1988.

[4] K. Birman, A. Schiper and P. Stephenson, "Lightweight Causal and Atomic Group Multicast", ACM Transactions on Computer Systems, vol. 9, no. 3, pp. 272-314, August 1991.

[5] H. Garcia-Molina and A. Spauster, "Ordered and Reliable Multicast Communication", ACM Transactions on Computer Systems, vol. 9, no. 3, pp. 242-272, August 1991.

[6] L.H. Ngoh, "Multicast Support for Group Communications", Computer Networks and ISDN Systems, vol. 22, pp. 165-178, 1991.

[7] M. Stonebreaker, "Concurency Control and Consistency of Multiple Copies of Data in Distributed INGRESS", IEEE Transactions on Software Engineering, pp. 188-194, May 1979.

[8] Marc Cohn, "High Speed Transport Protocol (HSTP) Specification", Contribution to ISO/IEC JTC1 SC6/WG4 on the High Speed Transport Protocol, September 1991.

[9] Protocol Engines Inc., "XTP Protocol Definition", Version 3.6, 11 January 1992.

[10] L. Aguilar et. al., "Architecture of a Multimedia Teleconferencing System", Proc. of ACM SIGCOMM, pp. 126-135, 1986.

[11] D. Cheriton, "VMTP: Versatile Message Transaction Protocol", RFC 1045, June 1988.

[12] J. Moulton, Proposed USA Contribution to SC6 on Multicast Transport Protocol, July 1992.

[13] M.D. Prycker, "Asynchronous Transfer Mode: Solution for Broadband ISDN", Ellis Horwood, 1991.

[14] P. Minet, "Performance Evaluation of GAM-T-103 Real Time Transfer Protocols", Proc. of IEEE Infocom 1989, Ottawa, April 1989.

[15] D. Clark, M. Lambert, L. Zhang, "NETBLT: A Bulk Data Transfer Protocol", Request For Comments, RFC 998, March 1987.

[16] R.W. Watson, "The Delta-t Transport Protocol: Features and Experience", Proc. of IFIP Workshop on Protocols for High Speed Networks, pp. 3-18, Zurich, 1989.

[17] R.W. Watson, "Gaining Efficiency in Transport Services by Appropriate Design and Implementation Choices", ACM Transactions on Computer Systems, vol. 5, no. 2, pp. 97-120, May 1987.

[18] W.A. Doeringer et. al., "A Survey of Light-Weight Transport Protocols for High-Speed Networks", IEEE Transactions on Commmunications, vol. 38, no. 11, pp. 2025-2039, November 1990.

[19] Protocol Engine Document 90-81, "Multicast Connection Oriented Services", 3 pages.

[20] Mc. Quillan et. al., "The New Routing Algorithm for the Arpanet", IEEE Transactions on Commmunications, vol. 28, no. 5, May 1980.
[21] S. Sakata, T. Ueda, "A Distributed Interoffice Mail System", IEEE Computer, vol. 13, no. 10, pp. 106-116, October 1985.
[22] S. Lin, D.J. Costello Jr., "Protocols and Techniques for Data Communication Networks", F.F. Kuo Ed., Prentice Hall, 1979.
[23] D. Cheriton, W. Zwaenepoel, "Distributed Process Groups in the V Kernel", ACM Transactions on Computer Systems, vol. 3, no. 2, pp. 77-107, May 1985.
[24] E.J. Berglund, D. Cheriton, "Amaze: a Multiplayer Computer Games", IEEE Software, vol. 2, no. 3, 1985.
[25] S. Deering, "Host Extension for IP Multicasting", RFC 1112, August 1989.

CONGESTION CONTROL MECHANISMS FOR ATM NETWORKS[1]

Douglas S. Holtsinger

Department of Electrical and Computer Engineering, and
Center for Communications and Signal Processing
North Carolina State University
Raleigh, NC 27695

Abstract

In this paper, we survey the congestion control mechanisms which have been proposed for ATM networks. Much discussion has ensued on whether preventive or reactive congestion control mechanisms are best suited for use in ATM networks. We argue that a single congestion control mechanism cannot be optimal for serving the needs of all connections, due to the heterogeneous nature of traffic sources and the service requirements in ATM networks. We focus on comparing the effectiveness of congestion control mechanisms for managing the resource needs of particular types of connections.

INTRODUCTION

The introduction of high-speed switching and transmission technologies has opened up new opportunities for the development of high-speed communication networks capable of providing a diverse array of communication services in an integrated fashion. Particular attention has been directed towards the development of Broadband Integrated Services Digital Network (B-ISDN) architectures for providing these communication services. The Asynchronous Transfer Mode (ATM) is the transport technique of choice for the B-ISDN as proposed by the CCITT and T1 standards committees [72].

The development of congestion control mechanisms for B-ISDN/ATM networks has been a challenging task, principally due to the large transmission capacities available from broadband media, along with the large propagation delays which are normally associated with the communication links. The large transmission-propagation delay product of the communication links between network nodes allows a communication link to deliver a large number of cells to a congested node before the network control elements can throttle the offending source [2]. In addition, the large propagation delays

[1]Supported in part by BellSouth, GTE Corporation, and NSF and DARPA under cooperative agreement NCR-8919038 with the Corporation for National Research Initiative.

Asynchronous Transfer Mode Networks, Edited by Y. Viniotis
and R.O. Onvural, Plenum Press, New York, 1993

may give rise to long periods between the onset of elevated traffic conditions at a node and the detection of these conditions by the appropriate network control elements [69].

Congestion control mechanisms are designed to ensure that each connection in the network receives a guaranteed quality-of-service (QOS), such as guarantees on throughput, cell loss, delay, and delay variance. The congestion control mechanisms for B-ISDN/ATM networks should be simple to accommodate the large link transmission speeds, and flexible enough to accommodate a wide variety of traffic patterns and QOS requirements [16].

There are two congestion control techniques which are traditionally considered: reactive and preventive congestion control [70]. In reactive (or closed-loop) congestion control, the network reacts to elevated network traffic conditions by taking active measures to ensure that congestion conditions do not arise. Preventive (or open-loop) congestion control seeks to prevent congestion conditions from arising through mechanisms designed to limit the amount of traffic which enters the network. Our purpose here is to review some of the preventive and reactive congestion control mechanisms which have been proposed for use with B-ISDN/ATM networks. We begin by reviewing preventive congestion control techniques, followed by the reactive congestion control mechanisms, and finally, we present our conclusions.

PREVENTIVE CONGESTION CONTROL

Connection Admission Control

A connection admission control (CAC) mechanism determines whether a call can be accepted into the network, and allocates network resources to the connection if the call is accepted [71]. Resources such as buffers and transmission capacity are typically allocated once during the setup of a connection [31]. We describe techniques for dynamically allocating resources during the call in an upcoming section.

If the network cannot accept the connection, then the CAC mechanism may either reject the call or it may negotiate the parameters of the traffic contract with the source. If the call is accepted, then the CAC mechanism determines the parameters of the policing mechanism, which is designed to ensure that the connection adheres to the limits of the traffic contract. Policing mechanisms are described in more detail in a later section. We can broadly classify CAC mechanisms into three categories: mechanisms based upon queueing models, virtual bandwidth models, and bufferless models.

Queueing Model Representation. A CAC mechanism based upon a queueing model assumes that the behavior of the traffic source can represented by a detailed statistical model, such as a Markov Modulated Bernoulli Process (MMBP). A queueing model is constructed which represents the network, and the CAC mechanism determines from the queueing model whether the network has the necessary buffer and transmission capacity to accommodate the traffic source (see for example, [11, 27]).

Modeling or simulating the entire network is usually computationally intensive, so we must often be satisfied with using simplified network models. Figure 1 shows a simplified network model of a single multiplexer queue, with cells arriving from a maximum of N sources. This model could also represent the buffered output multiplexer from a $N \times N$ crossbar switch.

These models are generally acknowledged to have serious practical limitations as CAC mechanisms, since obtaining estimates of the end-to-end network performance is non-trivial, and it becomes difficult to account for heterogeneous sources. Numerous approximations have been proposed to simplify the calculation of the performance metrics (see Hong [30] for an excellent summary).

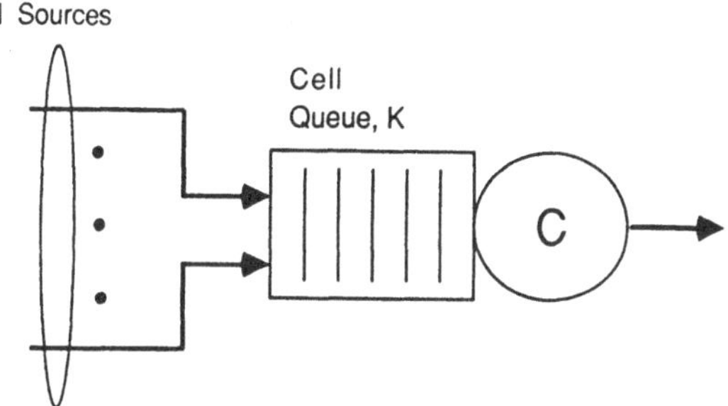

Figure 1. Simplified Network Model

A promising approach taken by Kurose [39] is to stochastically bound the distribution of the number of cells which can be generated by each source over various lengths of time, and then "pushing" these bounds through the network on a per-session basis. Bounds on the cell delay distribution can then be obtained once these bounds on the arrival process have been obtained.

Bufferless Model Representation. In a bufferless model representation, an upper bound on the cell loss probability of a finite-size cell queue multiplexer shown in figure 1 can be obtained by assuming that the cell queue size is zero. During each slot, C cells may be transmitted over the link, and up to N cells may arrive from the connections which share the link.

If the number of cells k which arrive during a slot is greater than C, then a maximum of $k - C$ cells will be dropped during the slot. An upper bound on the cell loss probability, $\hat{P}_{\text{cell loss}}$, can then be obtained as follows:

$$\hat{P}_{\text{cell loss}} = \frac{\sum_{k=0}^{N}[k - C]^{+}p(k)}{\sum_{k=0}^{N} k\, p(k)}$$

where $p(k)$ is the probability that k cells arrive during a slot from the total of N connections, and

$$[x]^{+} = \begin{cases} x, & x > 0 \\ 0, & \text{otherwise.} \end{cases}$$

These models allow good link utilization to be achieved if the source peak rate is much less than the transmission capacity of the output link (at least several orders of magnitude, see Suzuki et al [65]). Heterogeneous sources can easily be considered in this type of model [32, 33, 47], but it tends to become computationally intensive for a large numer of heterogeneous sources.

Saito and Shiomoto [61] proposed a call admission mechanism which relies on progressively refining an estimate of the cell arrival distribution through periodic measurement of the traffic characteristics. When a call arrives, the CAC mechanism uses the mean and peak cell arrival rate of the source to calculate an initial "worst-case" cell arrival distribution $\bar{p}_i = \{p_i(0), p_i(1), \ldots, p_i(N)\}$ of the individual source i. This "worst-case" distribution is convoluted with the present estimate of the aggregate cell arrival distribution $\bar{p}$ to determine whether the call can be accepted into the network using a bufferless model representation. Periodic measurements on the traffic stream are then performed to refine the estimate of the aggregate cell arrival distribution.

Virtual Bandwidth Representation. The concept of virtual bandwidth (also known as "equivalent bandwidth") is designed to capture the essential features of a source's traffic behavior using a simplified representation of the source's bandwidth requirements [10, 25]. Typically, the source's virtual bandwidth V is represented by a single number (or a vector of numbers), which would range from the mean transmission rate to the peak rate of the source, depending on the source characteristics.

The CAC mechanism accepts a new call into the network if the connection's virtual bandwidth V can be accommodated by all communication links between the source and destination of the connection. The virtual bandwidth capacity of a link is determined by subtracting the virtual bandwidth requirements of all connections passing through the link from the link capacity C. This approach has significant advantages over the queueing model representation, which requires the CAC mechanism to carry out a detailed queueing analysis at all links between the source and destination of the connection.

In the approach taken by Guérin, Ahmadi, and Naghshineh [24], the equivalent bandwidth (or capacity) $\hat{C}$ of a superposition of sources is represented by

$$\hat{C} = \min(\hat{C}_{(\mathrm{F})}, \hat{C}_{(\mathrm{S})})$$

where $\hat{C}_{(\mathrm{F})}$ is the sum of the equivalent capacities of all sources calculated using a fluid-flow model, and $\hat{C}_{(\mathrm{S})}$ is the sum of the equivalent capacities of all sources, calculated using a bufferless model representation, as described in the previous section.

The fluid-flow model grossly overestimates the equivalent capacity when the length of the source's burst period is long, and when a large number of sources are multiplexed together. The bufferless model representation is used under these conditions to give a more accurate estimate of the the equivalent capacity.

Another approach taken by Suzuki et al [65] is to allocate the peak transmission rate to sources which have a virtual bandwidth requirement close to their peak transmission rate. The remaining transmission capacity is divided among all other sources, with each source's bandwidth requirements estimated by a simple virtual bandwidth model. This improves the accuracy of the virtual bandwidth representation by minimizing the effect of statistical interference from high bandwidth sources.

Policing Mechanisms

A policing mechanism (also known as the Usage Parameter Control (UPC) in CCITT terminology [71, 73]) ensures that the traffic source conforms to the traffic contract which it has negotiated with the network at the time of the connection setup. As long as the traffic source complies with the limits of the traffic contract, the policing mechanism should remain transparent. If the source violates the limits of the traffic contract, then the policing mechanism should operate on the cell stream so as to prevent the source from inducing congestion conditions in the network.

Leaky bucket-type policing mechanisms. In a leaky bucket-type mechanism, cells are required to consume tokens before they pass through the leaky bucket [4, 5, 6, 29, 63, 64, 67]. Tokens are held in a token pool of fixed size, and tokens are periodically added to the token pool. Cells which arrive and find insufficient tokens in the token pool may be dropped, delayed, or "violation tagged", depending on the particular design. From the standpoint of cell loss, the leaky bucket can be shown to operate similarly to a finite capacity single server queue with a deterministic service time [15].

Bala, Cidon and Sohraby [3] proposed a leaky bucket-type policing mechanism which has two separate token pools for holding "red" and "green" tokens, generated at different rates. Packets wait in a single infinite size packet queue, and they consume a variable number of tokens of a single color, depending on the packet size. If the packet queue size is less than K, then the packet at the head of the queue waits for a sufficient number of "green" tokens to accumulate before departing. If the packet queue size is greater than or equal to K, then the packet may depart if there are a sufficient number of "red" tokens available. Packets which consume "red" tokens are dropped first if congestion conditions arise in the network.

Leaky bucket-type policing mechanisms are generally ineffective for policing both the mean rate and the burstiness of a traffic source simultaneously [9, 57]. Holtsinger [28] proposed the dual leaky bucket policing mechanism (see figure 2) to solve this problem. One token pool is configured to police the mean rate of the traffic source, and the other token pool is configured to police the burstiness of the source. The size of the token pool can be greatly reduced by imposing an upper limit on the burst length of the traffic source, and a lower limit on its silence period.

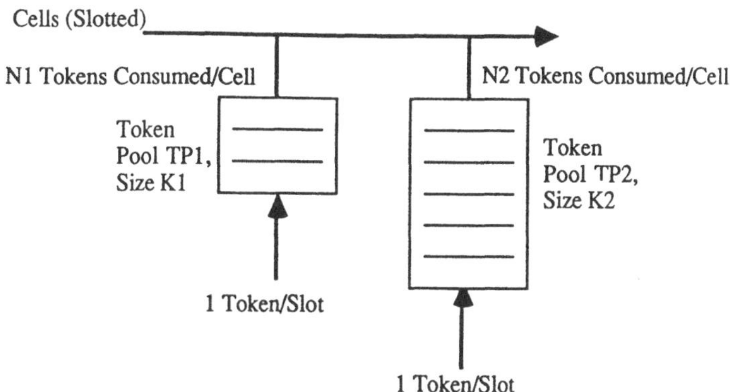

Figure 2. The dual leaky bucket policing mechanism

A leaky bucket proposed by Murata, Ohba, and Miyahara [48] uses a cell spacing algorithm to limit the peak transmission rate of the source. When a cell consumes tokens from the token pool, it is forced to wait P slots before being transmitted by a server, where P is the reciprocal of the peak transmission rate. A cell which arrives to find insufficient tokens in the token pool and no waiting cells is forced to wait A slots before being transmitted by a server (where A is the mean interarrival time of cells). If there are cells already waiting, then the arriving cell is enqueued to receive a service of A slots until the server clears the backlog of waiting cells.

Window Policing Mechanisms. Jumping window policing mechanisms operate by allowing the source to submit a maximum of N cells every T slots. Excess cells may either be dropped, delayed, or "violation-tagged", as in the leaky bucket. A leaky bucket-type mechanism with a token pool of size N and arrivals of N tokens every T slots operates identically to the jumping window mechanism. Moving window policing mechanisms accept a maximum of N cells during any non-overlapping interval of length T slots. In the triggered jumping window, the beginning of the window T is triggered by the first arriving cell [57].

The Exponentially Weighted Moving Average (EWMA) mechanism uses fixed time intervals like the jumping window, except that the maximum number of cells which the mechanism can accept during an interval of length T is a function of the mean number of accepted cells per interval (N) and an exponentially weighted sum of the number of accepted cells in preceding intervals. Rathgeb [57] found that the leaky bucket-type policing mechanism and the EWMA mechanism give superior performance over the jumping, moving, and triggered jumping window mechanisms.

Distributed Source Control (DSC), proposed by Ramamurthy and Dighe [55, 56], is a jumping window mechanism which allows W_s cells to pass into the network every T_s slots. Excess cells are held in an infinite size buffer at the UNI (see figure 3). DSC requires the use of a sliding window flow control mechanism at the transport level, with a window size of W_e. The maximum average throughput allowed by the window flow control mechanism is

$$\bar{\lambda} = \frac{W_e}{T_r}$$

where T_r is the round-trip propagation delay. The DSC parameters are determined such that the rate control mechanism does not constrain the maximum average throughput of the connection $\bar{\lambda}$ by setting

$$\frac{W_s}{T_s} = \frac{W_e}{T_r}$$

where typically $W_s < W_e$.

For a new call to be accepted, two constraints must be met at each node along the source-destination path. First, the sum of the bandwidths must be less than the link capacities, and secondly, the sum of the window sizes W_s for each call must be less than the buffer size at each node.

Another jumping window mechanism was proposed by Golestani [21], which is intended to be used in conjunction with a particular service discipline at nodes referred to as *stop-and-go* queueing. The jumping window rate control mechanism forces the traffic source to submit a maximum of rT bits into the network during every fixed length frame of length T. This smoothness property is preserved throughout the network by a service discipline known as *stop-and-go queueing.*

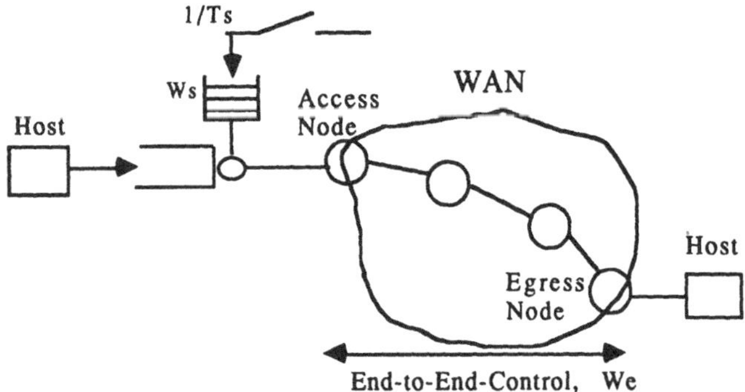

Figure 3. Distributed Source Control

Time is divided into frames of fixed length T, whereas the input and output links can have different frame boundaries which respect to each other. A packet which arrives to a node during a particular frame may not be transmitted until the beginning of the first departing frame after the arriving frame ends. The stop-and-go service discipline, along with the jumping window rate control mechanism, allows the cell jitter to be bounded by twice the duration of the frame length T. It is also shown that buffer overflow can be eliminated by dimensioning the buffer size as

$$B = 3C_l T$$

where C_l is the transmission capacity of the link. Extensions to this framing strategy allow for multiple frame sizes T and for provisioning of services which do not require stringent loss and delay guarantees [22, 23].

Rate Control and Traffic Shaping Mechanisms. Several other types of policing mechanisms have been proposed in the literature which act primarily to limit the source's traffic rate, or to "shape" and smooth the cell arrival process [15, 37, 45, 50, 59, 60]

A leaky bucket with a cell queue is often referred to as a traffic shaping mechanism (Eckberg, et al, [15]). Sidi et al [63] illustrated the tradeoff between the squared coefficient of variation of the interdeparture time and the cell waiting time when a leaky bucket is configured to shape the departure process. As the token pool size decreases, the burstiness of the departure process also decreases, while the mean cell waiting time increases.

The virtual clock mechanism has been proposed by Zhang [74] as a rate control mechanism to aid in the scheduling of cell transmissions. Each switching node along the source-destination path of a connection maintains a virtual clock v_i which is used for determining scheduling priorities with respect to other connections. The average transmission rate of connection i is represented by $\bar{\lambda}_i$. When a cell belonging to connection i arrives to a node, the virtual clock v_i is set to

$$\max(v_i + 1/\bar{\lambda}_i, \text{RealArrivalTime}),$$

and the arriving cell is time-stamped with the new value of the virtual clock. The cell with the minimum time-stamp value is always first to be scheduled for transmission. This mechanism allows sources which transmit faster than their average rate to still get scheduled if the resource is free, but the mechanism ensures that other connections receive fair access.

Weinrib and Wu [68] showed how the virtual clock operates in an "equivalent" manner to the leaky bucket policing mechanism in that the equations that their underlying state variables satisfy are closely related.

Tagging and Buffer Space Priority Mechanisms

A traffic source, or the UNI, may optionally choose to "tag" cells which represent "excess" traffic that the network may optionally drop if sufficient resources are not available for transporting the excess traffic [15, 42, 52]. Tagged cells can be dropped inside the network when congestion conditions arise at a network node, and the network node may assign lower buffer space priority to tagged cells.

Eckberg et al [16] suggested that sources such as video and voice could "pre-mark" non-essential cells with a violation tag, and to have the network carry these cells "at risk".

Tagging is often used in conjunction with buffer space priority mechanisms which give lower space priority to tagged cells [17, 38, 42]. Kröner [38] compared the effectiveness of two buffer priority mechanisms known as *push-out* and *partial buffer sharing* to a system with separate buffer space for each traffic class, and the one with complete buffer sharing and no priorities. In the push-out mechanism, an arriving high priority cell can replace a low priority ("tagged") cell in the buffer if there is no buffer space available upon arrival. In partial buffer sharing, an arriving cell of class i is dropped if there S_i or more buffer spaces occupied. The partial buffer scheme is less complex to implement than the push-out scheme, and it provides nearly comparable performance in terms of link utilization.

Fast Resource Allocation

Fast resource allocation protocols are designed to accommodate sources which cannot be efficiently statistically multiplexed, and sources whose characteristics are not known during call setup [7, 12, 13, 14]. These types of sources cannot be easily accommodated with open-loop congestion control mechanisms, such as policing mechanisms. In a fast resource allocation protocol, network resources are not reserved as in a CAC mechanism, but they are requested when the user has data to transmit. Network resources may be deallocated by the network during idle periods, or upon the user's request, but the connection always remains active for the duration of the call.

A protocol known as "short hold mode communication" was proposed by Ohnishi, Okada, and Noguchi [49] (see figure 4). In short hold mode, when the user finishes transmitting a burst of cells, the user sends a temporary resource clear request signal to the network which releases the resources held by the call. When the user wants to transmit a new burst of cells, the user sends a setup message to reallocate network resources to the call. The disadvantages are that the reallocation of network resources

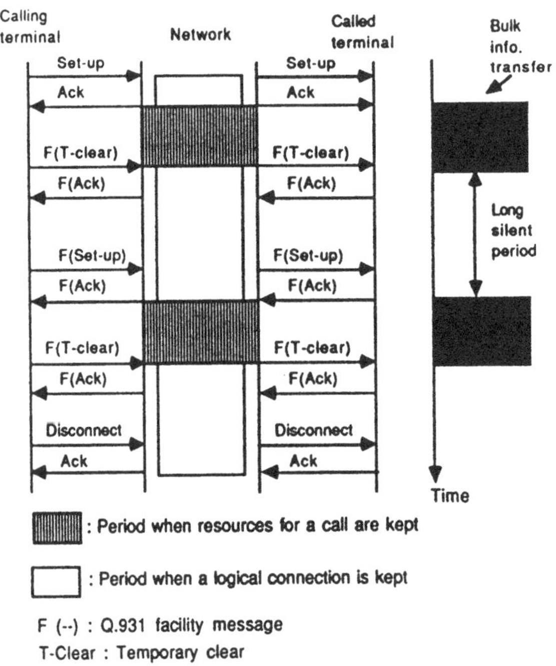

Figure 4. Short Hold Mode Communication

causes a significant round-trip propagation delay, and a user may not be able to clearly delineate the end of a burst, since bursts may be separated by very short intervals.

Boyer and Roberts proposed eliminating the round-trip propagation delay involved when reallocating network resources by allowing the source to transmit a burst of data immediately after a reservation request is submitted [8]. If the reservation request is denied, the cells belonging to the burst are simply discarded by the network. Presumably, the source would re-request network resources if the network rejected the initial reservation request.

Suzuki and Tobagi investigated a fast resource allocation scheme which relied upon routing individual bursts of the same connection through multiple source-destination paths and multiple links [66]. They found that their scheme significantly improved link utilization over a scheme which relied upon single-path and single-link routing.

REACTIVE CONGESTION CONTROL

Adaptive Window Flow Control

While window flow control is more concerned with ensuring that end-users are not overwhelmed with packets, it does have secondary effects on network congestion control. Window flow control mechanisms have been proposed which rely on adaptively adjusting the the window size in response to explicit network congestion indicators generated at intermediate network nodes [1, 40, 53, 54], or by adjusting window sizes in response to implicit congestion indicators, such as excessive cells delays or cell losses [35]. Another technique is to combine window flow control with an underlying rate control mechanism to keep users from submitting an entire window's worth of cells instantaneously [18, 41, 55, 56].

An adaptive sliding window flow control scheme was proposed by Pingali, Tipper, and Hammond [53] in which the window size is reduced to a minimum size $W_{\min}$ if the number of cells at any node is greater than an upper threshold UT, and the window is increased by one up to a maximum size $W_{\max}$ if all nodes along the path have a queue size less than a lower threshold LT. A similar scheme was proposed by Aboul-Magd, Gilbert, and Wernik [1], in which a node sets an explicit congestion notification (ECN) bit in the header field of cells passing in the forward direction when the the average queue length is above a fixed threshold. Upon receipt by the destination, the ECN bit is copied into the acknowledgement packet that returns to the source, which in turn is used at the source to adjust its window size.

Jain and Ramakrishnan [36] found that a multiplicative decrease in the window size (multiplying the window size by a factor r) upon the detection of congestion conditions and an additive increase in the window size during normal network conditions gave the best results with respect to fairness among users and optimality of user throughput. They also suggested that window sizes should be adjusted once every two round-trip delays to allow time for window adjustments to take effect (thus minimizing oscillations), and that the network congestion indicators from the last round-trip delay should be used for adjusting window sizes.

With sliding window flow control [62], the buffer size at intermediate nodes must be sufficiently large to hold a window's worth of cells for each connection which traverses the node in order to prevent buffer overflow, which can become prohibitively expensive.

Since users who share the network with other users do not need a full-speed round-trip window size to fully utilize network bandwidth, Hahne, Kalmanek and Morgan [26] proposed allocating a window size less than the round-trip window to new virtual circuit requests, but large enough to prevent the virtual circuit from becoming throughput-limited. By allocating window sizes to new virtual circuit requests as a decreasing function of the number of active virtual circuits, they found that the total buffer space at a node grew on the order of log N, where N is the number of virtual circuit users. To maintain fairness in the presence of different window sizes for each user, each virtual circuit is assigned separate buffer space in a node, and the node transmits cells in a round-robin fashion among all virtual circuits.

Explicit Congestion Notification

Explicit congestion notification (ECN) is a mechanism whereby intermediate nodes signal elevated network conditions by either sending control messages (or "choke" packets) back to the source [20, 34, 51], or setting bits in the header of cells passing in the forward direction [1, 43]. These bits are then returned to the source in the form of acknowledgements.

The node may also set the bits in the headers of cells passing in the reverse direction, which has the advantage of reaching the source faster, but the forward and reverse traffic may not always be related [36]. Setting the bits in the header of a cell instead of using separate control messages reduces the signaling overhead.

As we saw in the previous section, ECN may be used with window flow control to reduce the window size at the source during network congestion conditions. Alternatively, ECN may be combined with rate-controlled servers at the source, such as leaky bucket-type mechanisms, to reduce the source traffic rate during congestion conditions.

Figure 5 shows a Explicit Forward Congestion Notification (ECFN) mechanism proposed by Makrucki [43]. A user submits packets to a variable-rate server at the source, whose rate is dynamically controlled by messages sent back from the destination. A leaky bucket type mechanism at the destination filters the EFCN-marked stream of cells and sends a "back-off" message to the source server when it overflows.

While the large propagation delays in B-ISDN/ATM networks inhibits the ability of ECFN mechanisms to react quickly to elevated network traffic conditions, it has been noted that EFCN has some effectiveness over the long-term even when operating with large delays [44], and that they are most effective when they have congestion-anticipation information to overcome the propagation delays.

Combining Reactive and Preventive Congestion Control

Mukherjee, Landweber, and Faber [46] suggested a two-level congestion control strategy for data traffic (figure 6). Open-loop rate controls are used to police and meter user traffic to protect the network from short-term bursts of traffic, and a high-level closed-

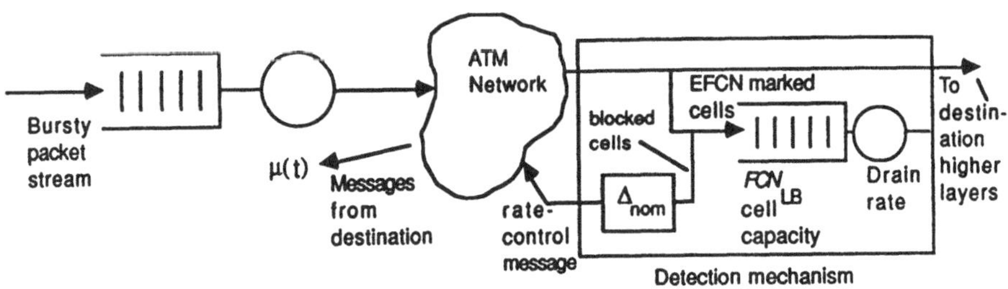

Figure 5. Explicit Forward Congestion Notification Scheme

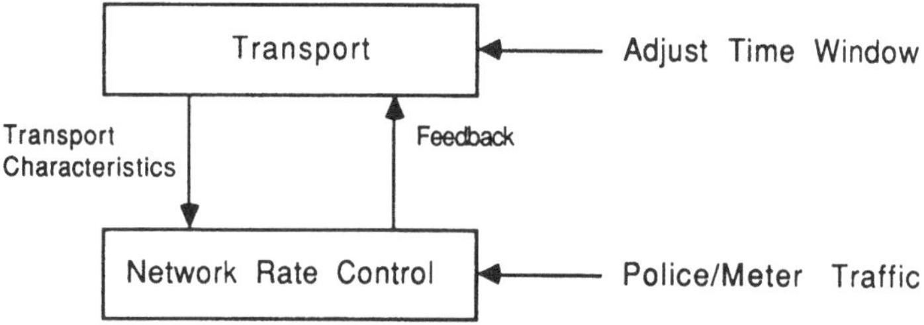

Figure 6. Two level control for data traffic

loop control mechanism provides some flexibility to deal with longer-term congestion conditions. They proposed a generalized virtual clock to police sources at the low-level called "Pulse Scheduling", and a jumping window mechanism with dynamically-adjustable window sizes at the high-level called "Dynamic Time Windows" (DTW). Pulse scheduling is used for giving switching priorities to cells, and DTW is an end-to-end transport level congestion avoidance mechanism based upon admission control. The two mechanisms share traffic parameters, but they act independently.

CONCLUSIONS

Due to the heterogeneous nature of traffic and service requirements in B-ISDN/ATM networks, it has become apparent that a single congestion control mechanism cannot be optimal for serving the needs of all sources. Policing and CAC mechanisms seem ill-suited for handling sources which generate non-stationary, unpredictable, or extremely bursty traffic, such as LAN-to-LAN traffic [19] or compressed video sources [58]. Fast resource allocation protocols seem better equipped to deal with extremely bursty sources and high peak bit rate sources, as long as connections can tolerate burst blocking or relatively long delays.

Violation tagging of non-essential cells appears to be a promising approach for carrying compressed video sources, but it's not clear whether CAC mechanisms, which generally assume stationary source characteristics, can efficiently manage non-stationary video sources. Traffic shaping mechanisms, like policing mechanisms, rely on the knowledge of the source characteristics, and so they could induce large delays and cell losses when used with extremely unpredictable or non-stationary traffic.

Since reactive congestion control schemes such as ECN generally rely on adaptive window flow control, or variable rate control placed at the UNI, their applicability is limited to sources which can be throttled and efficiently statistically multiplexed with other sources. Their applicability then, generally, overlaps with policing and traffic shaping mechanisms, but they are seen as more aggressive approaches towards increasing network utilization. Reactive schemes often share the same rate-control and traffic shaping functions at the UNI with policing mechanisms, and hence they can be viewed as generalizations of policing mechanisms. The question that remains to be answered is whether their performance gains (if any) can justify their increased complexity.

References

[1] O. Aboul-Magd, H. Gilbert, and M. Wernik. Flow and congestion control for broadband packet networks. In *Proc. 13th International Teletraffic Conference*, pages 853–858, 1991.

[2] J.J. Bae and T. Suda. Survey of traffic control schemes and protocols in ATM networks. *Proceedings of the IEEE*, 79(2):170–189, February 1991.

[3] K. Bala, I. Cidon, and K. Sohraby. Congestion control for high speed packet switched networks. In *IEEE Infocom*, pages 520–526, 1990.

[4] A.W. Berger. Performance analysis of a rate-control throttle where tokens and jobs queue. *IEEE Journal on Selected Areas in Communications*, 9(2):165–170, February 1991.

[5] A.W. Berger, A.E. Eckberg, T. Hou, and D.M. Lucantoni. Performance characterizations of traffic monitoring, and associated control, mechanisms for broadband "packet" networks. In *IEEE Globecom*, pages 350–354, 1990.

[6] A.D. Bovopolous. Performance evaluation of a traffic control mechanism for ATM networks. In *IEEE Infocom*, pages 469–478, 1992.

[7] P. Boyer. A congestion control for the ATM. In *Proc. 13th International Teletraffic Conference*, October 1990.

[8] P.E. Boyer and J.W. Roberts. Traffic control for stepwise VBR connections in an ATM network. In *Integrated Broadband Communications: views from RACE*, pages 391-398, 1991.

[9] M. Butto', E. Cavallero, and A. Tonietti. Effectiveness of the leaky bucket policing mechanism in ATM networks. *IEEE Journal on Selected Areas in Communications*, 9(3):335–342, April 1991.

[10] M. Decina and T. Toniatti. On bandwidth allocation to bursty virtual connections in ATM networks. In *Proc. of the ICC*, pages 844–851, 1990.

[11] L. Dittmann and S.B. Jacobsen. Statistical multiplexing of identical bursty sources in an ATM network. In *IEEE Globecom*, pages 1293–1297, 1988.

[12] B. Doshi. Performance of in-call buffer/window allocation schemes for short intermittent file transfers over broadband packet networks. In *IEEE Infocom*, pages 2463–2471, 1992.

[13] B. Doshi, S. Dravida, and G. Ramamurthy. Memory, bandwidth, processing and fairness considerations in real time congestion control for broadband networks. In *Proc. 13th International Teletraffic Conference*, pages 143–149, 1991.

[14] B. Doshi and H. Heffes. Performance of an adaptive buffer/window allocation scheme for long file transfers over wide area high speed packet networks. *IEEE Journal on Selected Areas in Communications*, 9(3):325–334, April 1991.

[15] A.E. Eckberg, D.T. Luan, and D.M. Lucantoni. Bandwidth management: A congestion control strategy for broadband packet networks – characterizing the throughput-burstiness filter. In *Int. Teletraffic Cong. Specialist Sem.*, Adelaide, Australia, September 1989.

[16] A.E. Eckberg, D.T. Luan, and D.M. Lucantoni. Meeting the challenge: congestion and flow control strategies for broadband information transport. In *IEEE Globecom*, pages 1769–1773, 1989.

[17] A.I. Elwalid and D. Mitra. Fluid models for the analysis and design of statistical multiplexing with loss priorities on multiple classes of bursty traffic. In *IEEE Infocom*, pages 415–425, 1992.

[18] K.W. Fendwick, D. Mitra, I. Mitrani, M.A. Rodrigues, J.B. Seery, and A. Weiss. An approach to high-performance, high-speed data networks. *IEEE Communications Magazine*, pages 74–82, October 1991.

[19] H.J. Fowler and W.E. Leland. Local area network traffic characteristics, with implications for broadband network congestion management. *IEEE Journal on Selected Areas in Communications*, 9(7):1139–1149, September 1991.

[20] A. Gersht and K.J. Lee. A congestion control framework for ATM networks. *IEEE Journal on Selected Areas in Communications*, 9(7):1119–1129, September 1991.

[21] S.J. Golestani. Congestion-free transmission of real-time traffic in packet networks. In *IEEE Infocom*, pages 527–536, June 1990.

[22] S.J. Golestani. Duration-limited statistical multiplexing of delay-sensitive traffic in packet networks. In *IEEE Infocom*, pages 323–332, Miami, Fl., April 1991.

[23] S.J. Golestani. A framing strategy for congestion management. *IEEE Journal on Selected Areas in Communications*, 9(7):1064–1077, Sept. 1991.

[24] R. Guérin, H. Ahmadi, and M. Naghshineh. Equivalent capacity and its application to bandwidth allocation in high-speed networks. *IEEE Journal on Selected Areas in Communications*, 9(7):968–981, Sept. 1991.

[25] R. Guérin and L. Gün. A unified approach to bandwidth allocation and access control in fast packet-switched networks. In *IEEE Infocom*, pages 1–12, 1992.

[26] E.L. Hahne, C.R. Kalmanek, and S.P. Morgan. Fairness and congestion control on a large ATM network with dynamically adjusted windows. In *Proc. 13th International Teletraffic Conference*, pages 861–872, 1991.

[27] M. Hirano and N. Watanabe. Characteristics of a cell multiplexer for bursty ATM traffic. In *Proc. of the ICC*, pages 399–403, 1989.

[28] D.S. Holtsinger. *Performance analysis of leaky bucket policing mechanisms for high-speed networks.* PhD thesis, Department of Electrical and Computer Engineering, North Carolina State University, 1992.

[29] D.S. Holtsinger and H.G. Perros. Performance analysis of leaky bucket policing mechanisms. Technical report, Department of Computer Science, North Carolina State University, 1991.

[30] S.W. Hong. *Modelling and analysis of the shared buffer ATM switch architecture for broadband integrated services digital networks.* PhD thesis, Department of Computer Science, North Carolina State University, 1992.

[31] J.Y. Hui. Resource allocation for broadband networks. *IEEE Journal on Selected Areas in Communications*, 6(9), 1988.

[32] B. Jabbari. A bandwidth allocation technique for high speed networks. In *IEEE Globecom*, pages 355–359, 1990.

[33] B. Jabbari and F. Yegenoglu. An upper bound for cell loss probability of bursty sources in broadband packet networks. In *Proc. of the ICC*, pages 699–703, 1991.

[34] S. Jagannath and I. Viniotis. A novel architecture and flow control scheme for private ATM networks. In *Proc. of TriComm '92*, pages 135–155, Raleigh, NC, February 1992.

[35] R. Jain. A timeout-based congestion control scheme for window flow-controlled networks. *IEEE Journal on Selected Areas in Communications*, 4(7):1162–1167, October 1986.

[36] R. Jain and K.K. Ramakrishnan. Congestion avoidance in computer networks with a connectionless network layer: concepts, goals and methodology. In *Computer Networking Symposium*, pages 134–143, Washington DC, April 1988.

[37] A. Jalali and L.G. Mason. Open loop schemes for network congestion control. In *Proc. of the ICC*, pages 199–203, 1991.

[38] H. Kröner. Comparative performance study of space priority mechanisms for ATM networks. In *IEEE Infocom*, pages 1136–1143, 1990.

[39] J. Kurose. On computing per-session performance bounds in high-speed multi-hop computer networks. *Performance evaluation review*, 20(1):128–139, June 1992.

[40] W.E. Leland. Window-based congestion management in broadband ATM networks. In *IEEE Globecom*, pages 1794–1800, 1989.

[41] D.T. Luan and D.M. Lucantoni. The effect of bandwidth management on the performance of a window-based flow control. *AT&T Technical Journal*, pages 17–26, September/October 1988.

[42] B. Mackrucki. A study of source traffic management and buffer allocation in ATM networks. In *International Teletraffic Congress Seventh Specialist Seminar on Broadband Technologies*, Morristown, October 1990.

[43] B. Makrucki. On the performance of submitting excess traffic to ATM networks. In *IEEE Globecom*, Phoenix, 1991.

[44] B. Makrucki. Explicit forward congestion notification in ATM networks. In *Proc. of TriComm '92*, pages 113–134, Raleigh, NC, February 1992.

[45] J.A.S. Monteiro, M. Gerla, and L. Fratta. Input rate control for ATM networks. In *Proc. 13th International Teletraffic Conference*, pages 117–122, 1991.

[46] A. Mukherjee, L.H. Landweber, and T. Faber. Dynamic time windows and generalized virtual clock: combined closed-loop/open-loop congestion control. In *IEEE Infocom*, pages 322–332, 1992.

[47] T. Murase, H. Suzuki, and T. Takeuchi. A call admission control for ATM networks based on individual multiplexed traffic characteristics. In *Proc. of the ICC*, pages 193–198, 1991.

[48] M. Murata, Y. Ohba, and H. Miyahara. Analysis of flow enforcement algorithm for bursty traffic in ATM networks. In *IEEE Infocom*, pages 2453–2462, 1992.

[49] H. Ohnishi, T. Okada, and K. Noguchi. Flow control schemes and delay/loss tradeoff in ATM networks. *IEEE Journal on Selected Areas in Communications*, 6(9):1609–1616, December 1988.

[50] T. Ott, T.V. Lakshman, and A. Tabatabai. A scheme for smoothing delay-sensitive traffic offered to ATM networks. In *IEEE Infocom*, pages 776–785, 1992.

[51] A. Periyannan. *A reactive congestion control scheme for gateways on high-speed networks.* M.S. Thesis, Department of Computer Science, North Carolina State University, 1992.

[52] D.W. Petr and V.S. Frost. Optimal packet discarding: an ATM-oriented analysis model and initial results. In *IEEE Infocom*, pages 537–542, 1990.

[53] S. Pingali, D. Tipper, and J. Hammond. The performanc of adaptive window flow controls in a dynamic load environment. In *IEEE Infocom*, pages 55–62, 1990.

[54] K.K. Ramakrishnan. A binary feedback scheme for congestion avoidance in computer networks with a connectionless network layer. In *Proc. SIGCOMM*, pages 303–313, August 1988.

[55] G. Ramamurthy and R.S. Dighe. Distributed source control: A network access control for integrated broadband packet networks. In *IEEE Infocom*, June 1990.

[56] G. Ramamurthy and R.S. Dighe. Distributed source control: A network access control for integrated broadband packet networks. *IEEE Journal on Selected Areas in Communications*, 9(7):990–1001, September 1991.

[57] E.P. Rathgeb. Modeling and performance comparison of policing mechanisms for ATM networks. *IEEE Journal on Selected Areas in Communications*, 9(3):325–334, April 1991.

[58] E.P. Rathgeb. Policing of realistic VBR video traffic – a case study. In *Proc. of IFIP TC6 workshop on broadband communications*, pages 287–300, Estoril, Portugal, January 1992.

[59] G. Rigolio and L. Fratta. Input rate regulation and bandwidth assignment in ATM networks: an integrated approach. In *Proc. 13th International Teletraffic Conference*, pages 123–128, 1991.

[60] G. Rigolio, L. Verri, and L. Fratta. Source control and shaping in ATM networks. In *IEEE Globecom*, pages 276–280, 1991.

[61] H. Saito and K. Shiomoto. Dynamic call admission control in ATM networks. *IEEE Journal on Selected Areas in Communications*, 9(7):982–989, September 1991.

[62] M. Schwartz. *Telecommunications networks: protocols, modeling and analysis.* Addison Wesley, 1987.

[63] M. Sidi, W. Liu, I. Cidon, and I. Gopal. Congestion control through input rate regulation. In *IEEE Globecom*, pages 49.2.1–49.2.5, 1989.

[64] K. Sohraby and M. Sidi. On the performance of bursty and correlated sources subject to leaky bucket rate-based access control schemes. In *IEEE Globecom*, pages 426–434, 1990.

[65] H. Suzuki, T. Murase, S. Sato, and T. Takeuchi. A burst traffic control strategy for ATM networks. In *IEEE Globecom*, pages 874–878, 1990.

[66] H. Suzuki and F.A. Tobagi. Fast bandwidth reservation scheme with multi-link and multi-path routing in ATM networks. In *IEEE Infocom*, pages 2233–2240, 1992.

[67] J.S. Turner. New directions in communications (or which way to the information age?). *IEEE Communications Magazine*, 24(10):8–15, October 1986.

[68] A. Weinrib and L.T. Wu. Virtual clocks and leaky buckets: flow control protocols for high-speed networks. In *IFIP WG 6.1/WG 6.4 Second International workshop on Protocols for high-speed networks*, pages 3–15, Palo Alto, CA, November 1990.

[69] G.M. Woodruff and Rungroj Kositpaiboon. Multimedia traffic management principles for guaranteed ATM network performance. *IEEE Journal on Selected Areas in Communications*, 8(3):437–446, April 1990.

[70] G.M. Woodruff, R.G.H. Rogers, and P.S. Richards. A congestion control framework for high-speed integrated packetized transport. In *IEEE Globecom*, pages 7.1.1–7.1.5, 1988.

[71] CCITT Study Group XVIII. Draft recommendation I.311, B-ISDN general network aspects. May 1990.

[72] CCITT Study Group XVIII. Draft recommendations. May 1990.

[73] CCITT Study Group XVIII. Annex to draft recommendation I.311, section 4, B-ISDN general network aspects. June 1991.

[74] L. Zhang. Virtual clock: A new traffic control algorithm for packet switching networks. *ACM Transactions on Computer Systems*, 9:101–124, Nov. 1991.

HIGHLY-BURSTY SOURCES AND THEIR ADMISSION CONTROL IN ATM NETWORKS

Khosrow Sohraby
IBM T.J. Watson Research Center
P.O. Box 704
Yorktown Heights, N.Y. 10598

Abstract

In this paper we focus on highly-bursty heterogeneous sources, their heavy traffic multiplexing behavior and their admission control in ATM networks.

The sources are characterized by their peak rate, utilization and the first two moments of their burst period. A heterogeneous statistical multiplexer fed with number of such sources is considered and its cell occupancy is analyzed.

We obtain a simple characterization of the tail of cell occupancy distribution of the multiplexer in the heavy traffic. This characterization provides a simple metric for call admission control in ATM networks where the network resources are shared by different sources with highly different characteristics. The proposed metric for call admission control is ideal for real time operation in ATM environment.

Introduction and Summary

Future high-speed networks are expected to support various traffic types with different characteristics and grade-of-service requirements such as data, voice, image, and video. The traffic which is generated from these services is substantially different in their characteristics. Understanding the impact of different traffic types on the performance is crucial for a successful and efficient design of such networks.

A call admission control policy determines whether a new connection can be accepted or not. The controller has to make sure that the grade-of-service (say cell loss probability) of all the connections in progress and also the newly accepted connection can be guaranteed. Therefore, it is natural to define a metric for each link in the network and at the connection set up time, the call admission controller examines the metric to decide if the call can be routed over the link. Obviously, each time a connection is accepted or taken down, the metric should be updated by taking into account the connection characteristics, e.g., its peak rate, utilization, etc. The reader is referred to [5, 6, 7, 13] and many references therein.

In this paper we generalize the results in [11] where the asymptotic behavior of a heterogeneous multiplexer fed by Binary Markov Sources was considered. This paper

Asynchronous Transfer Mode Networks, Edited by Y. Viniotis
and R.O. Onvural, Plenum Press, New York, 1993

focuses on highly-bursty sources where the burst period may have (an arbitrary) large variance. Our study shows that ignoring the burst size variability may lead to considerable error in estimating the cell occupancy (or cell loss probability) of the statistical multiplexer. We are concerned with a simple and effective scheme for the admission control of highly bursty sources. We will derive a metric for the call admission control and provide a simple algorithm for its real time implementation.

A high-speed link (or more precisely a Virtual Path, VP) in an ATM environment is modeled as a heterogeneous multiplexer fed by number of connections (or Virtual Circuits, VC). The queueing analysis of the multiplexer is carried out in discrete time which is representative of ATM environment at the cell level. Our results are general and can be readily used in other cases as well.

Using spectral decomposition method and heavy traffic analysis, we show that the tail behavior of the queue length distribution has a simple characterization. This characterization uses only each source's average utilization and the first two moments of the burst period and provides a simple approximation on the queueing behavior of the multiplexer, where the impact of each source is quite evident. We will provide simple and effective approximations which can be evaluated on the "back of an envelope" by a design engineer for buffer sizing, and a simple procedure for call admission control in high-speed environment which can effectively operate in real time.

In the past, researchers have used fluid flow models to address homogeneous and heterogeneous multiplexers (see for example [1]). However, most of the analysis is reported for the simple homogeneous case. Our work addresses the discrete nature of the problem directly and we give a simple analytical results on the tail behavior in general heterogeneous case.

In what follows, we present the main results and the summary of the paper. We assume that a link (or a multiplexer) is fed by K connections, each characterized by its peak rate, utilization, and the first two moments of its burst period. We denote the ratio of the link peak rate to the peak rate of connection i by R_i, its utilization by $\frac{\rho_i}{R_i}$, its average burst size by T_i and finally its squared coefficient of variation of the burst size by C_i^2. Therefore, a connection i is characterized by the quadruplet $(\rho_i, T_i, C_i^2, R_i)$. In situations where the squared coefficient of variation of the burst period is not known or cannot be estimated, we may assume that it is equal to one.

An immediate application of our results is in buffer sizing of a heterogeneous multiplexer fed by bursty arrivals. The simplest approximation on the queue length distribution is given by

$$\Pr(q > i) \approx \rho \left[1 + \frac{2(1-\rho)}{\sum_{i=1}^{K} \frac{\rho_i}{R_i}(1-\rho_i)^2(1+C_i^2)T_i}\right]^{-i}, \quad \rho \stackrel{\Delta}{=} \sum_{i=1}^{K} \rho_i \tag{1}$$

which can be easily calculated. Some discussion is in order here. First, this result is most accurate when the multiplexer is operating in the heavy traffic. Second, it is based on infinite (large) buffer assumption. Both of the above assumptions are expected to *overestimate* the actual buffer occupancy. Regarding the first assumption, we comment that usually the buffers are engineered for the heavy traffic case. For the second assumption, we comment that if the buffer is large and small probability of loss is desired (e.g., ATM networks), it is usually a good approximation to the finite buffer case.

The second and the main application of our simple result is in call admission control in high-speed networks. Modeling an ATM link (or a VC) as a heterogeneous multiplexer is common [5, 6, 7]. As discussed, it is desirable to define a simple and effective metric which takes into account a connection characteristic and based on its required grade-of-service, it indicates if that connection can be accepted. In this setting, for a given buffer size B (in terms of number of cells), a connection is accepted if the resulting loss probability (in our heterogeneous multiplexer) is less than a desired small number say ϵ. We have the following metric for the call admission control. The controller does NOT accept a connection unless $\rho\beta^B < \epsilon$, or the following is true:

$$Ln(\rho) - BLn(\left[1 + \frac{2(1-\rho)}{\sum_{i=1} \frac{\rho_i}{R_i}(1-\rho_i)^2(1+C_i^2)T_i}\right]) < Ln(\epsilon) \tag{2}$$

where Ln denotes the natural logarithm.

Somewhat even simpler metric for the call admission control policy may be given if we approximate the above result assuming that $\rho \approx 1$. We get the following simple result ($\kappa \stackrel{\Delta}{=} - Ln(\epsilon)$)

$$(\frac{\kappa}{1-\rho} - 1)\sum_{i=1} \frac{\rho_i}{R_i}(1-\rho_i)^2(1+C_i^2)T_i < 2B \tag{3}$$

Therefore, in either (38) or (39) we only have to update only two parameters, namely, the total utilization $\rho \stackrel{\Delta}{=} \sum_{i=1} \rho_i$, and the quantity $\sum_{i=1} \frac{\rho_i}{R_i}(1-\rho_i)^2(1+C_i^2)T_i$ each time a connection is taken down. To see if a new connection can be accepted, the two parameters are calculated to see if the above inequality will still hold, if it does, the connection is accepted, and the two parameters are accordingly updated.

Source Traffic Model and Formulation

In this section, we describe the environment and the mathematical model for the arrival process. Time is assumed slotted and the cell arrival process in each slot is governed by a homogeneous finite-state, aperiodic discrete-time Markov chain, called the *modulating Markov chain*, where transitions between states of the chain take place only at the slot boundaries. Denote S_n as the state of the modulating chain in slot n and p_{ij} for $1 \leq i,j \leq N$, as the transition probability of this chain, i.e, $p_{ij} = \Pr[S_{n+1} = j | S_n = i]$. Given the chain is in state i in slot n and makes a transition to state j in slot $n+1$, (which happens with probability $p_{i,j}$), the p.g.f of A_{n+1}, the number of cells arriving in slot $n+1$ is denoted by $a_{i,j}(z)$. We have

$$E[z^{A_{n+1}} \mathbf{1}\ (S_{n+1} = j) | S_n = i] \stackrel{\Delta}{=} a_{ij}(z) \tag{4}$$

where the indicator function $\mathbf{1}\ (E)$ is equal to one if the event E is true and equal to zero otherwise. It should be noted that in general, the number of cells arriving in a slot depends on the state of the modulating chain in both current and previous slot. This is a minor generalization of the arrival processes in discrete time considered by [12] and [11].

An important matrix which plays a major rule in the queueing analysis with the above arrival process is the NXN *probability generating matrix* (p.g.m) of the arrival process denoted by $\mathbf{A}(z)$ defined by

$$\mathbf{A}(z) \stackrel{\Delta}{=} \begin{bmatrix} a_{11}(z) & a_{12}(z) & \cdots & a_{1N}(z) \\ a_{21}(z) & a_{22}(z) & \cdots & a_{2N}(z) \\ \vdots & & & \vdots \\ a_{N1}(z) & a_{N2}(z) & \cdots & a_{NN}(z) \end{bmatrix} \tag{5}$$

where the matrix $\mathbf{A}(1)$ is simply the probability transition matrix of the modulating chain.

Now Consider a single server queue in discrete time with a Markovian cell arrival process governed by the p.g.m $\mathbf{A}(z)$. Denote Q_n as the number of cells in the queue in slot n. We assume that in any slot a maximum of one cell can be transmitted in any slot. We have the following evolution equation (infinite buffer size assumed)

$$Q_{n+1} = (Q_n - 1)^+ + A_{n+1} \tag{6}$$

where A_n is the total number of cells arriving in slot n with the p.g.m described above.

It is clear that the doublet (Q_n, S_n) is a Markov and has a transition probability matrix P of $M/G/1$ type [8]. We have

$$\mathrm{P} = \begin{bmatrix} \mathbf{A}_0 & \mathbf{A}_1 & \mathbf{A}_2 & \mathbf{A}_3 & \cdots \\ \mathbf{A}_0 & \mathbf{A}_1 & \mathbf{A}_2 & \mathbf{A}_3 & \cdots \\ 0 & \mathbf{A}_0 & \mathbf{A}_1 & \mathbf{A}_2 & \cdots \\ 0 & 0 & \mathbf{A}_0 & \mathbf{A}_1 & \cdots \\ \cdot & \cdot & \cdot & \cdot & \cdots \\ \cdot & \cdot & \cdot & \cdot & \cdots \end{bmatrix} \tag{7}$$

where the NXN matrices $\mathbf{A}_i$, $i \geq 0$ are the Taylor series expansion of the p.g.m $\mathbf{A}(z)$, i.e.,

$$\mathbf{A}(z) = \sum_{i=0}^{\infty} \mathbf{A}_i z^i \tag{8}$$

The above formulation of the arrival process can be readily used in modeling the superposition of heterogeneous sources each having its own probability generating matrix. Assume that the cell arrival process consists of superposition of K arrival processes independent from each other. Let $\mathbf{A}^{(i)}(z)$ denote the p.g.m of the ith arrival process, then it can be easily shown that the p.g.m. of the superposition process is simply the kronocker product of the individual processes [8]. We have

$$\mathbf{A}(z) = \mathbf{A}^{(1)}(z) \otimes \mathbf{A}^{(2)}(z) \otimes \dots\dots \otimes \mathbf{A}^{(K)}(z) \tag{9}$$

which has the dimension NXN, where $N = \prod_{i=1}^{K} N_i$ assuming that the p.g.m $\mathbf{A}^{(i)}(z)$ of source i has dimension $N_i X N_i$.

Now for a general source with p.g.m $\mathbf{A}(z)$, assuming that the matrix $\mathbf{A}(1)$ is irreducible and aperiodic, denote ρ as the average number of arrivals per slot, then the chain P has a stationary solution if $\rho < 1$. It can be easily shown that $\rho = \pi\mathbf{A}(1)e$ where the $1XN$ vector π is the stationary distribution of the underlying modulating chain $A(1)$ satisfying $\pi = \pi A(1)$ and the $NX1$ vector e is the unit vector with all its elements equal to 1.

Define

$$Q_j(z) \triangleq \lim_{n \to \infty} \sum_{i=0}^{\infty} \Pr(Q_n = i, S_n = j) z^i \tag{10}$$

$$\overline{Q}(z) \triangleq [Q_1(z), Q_2(z), \dots\dots, Q_N(z)] \tag{11}$$

Following the standard z transform arguments, we have

$$\overline{Q}(z)[z\mathbf{I} - \mathbf{A}(z)] = (z-1)\overline{Q}(0)\mathbf{A}(z) \tag{12}$$

In principle, for a stable system the boundary vector $\overline{Q}(0)$ is uniquely determined by the N zeros (counting multiplicities) of the determinant $\Delta(z) = \mid z\mathbf{I} - \mathbf{A}(z) \mid$ inside and on the unit circle. However, if N the dimension of $\mathbf{A}(z)$ is large, the exact calculation of queue length distribution is by no means a trivial task. The main difficulty is due to an accurate and efficient calculation of all the zeros of the determinant inside the unit circle and forming the N linear equations for solving the boundary vector $\overline{Q}(0)$. To circumvent this problem, Neuts has introduced an iterative method using matrix analytic approach [8], which usually for large N requires an extensive amount of calculations especially in the heavy traffic.

It can be shown that in the special case which only one column of the matrix $\mathbf{A}(0)$ (assumed to be the first column without loss of generality) is nonzero, the vector $\overline{Q}(0)$ is readily available without any calculation and we have

$$\overline{Q}(0) = (1-\rho)(1,0,0,.....,0) \tag{13}$$

Example of the above structure may be found in [4] and [10]. Even in the above case which $\overline{Q}(z)$ is completely known, the exact distribution of the queue length requires extensive calculation. Our approach in this paper is to provide a *simple and effective* approximation of the *queue length distribution* which can be easily applied to a general arrival process as described above. We mainly focus in the heavy traffic case which is usually of main interest in performance analysis and on the tail behavior of the queue length distribution which dictates the high percentiles and could be used in approximating very small loss probabilities.

It is generally known that if the steady-state queue length distribution has infinite support, it exhibits a geometric tail behavior which is characterized by the smallest root outside the unit circle of the determinant $\mid z\mathbf{I} - \mathbf{A}(z) \mid$. Denote this root by $z^\star$, and define $\beta \triangleq \frac{1}{z^\star}$ then for sufficiently large i, we have

$$\Pr(q > i) \approx \alpha\beta^i \tag{14}$$

where $\Pr(q > i) \triangleq \lim_{n\to\infty} \Pr(Q_n > i)$. In many queueing situations of interest, this geometric tail behavior is usually apparent even following moderate percentiles especially if the system is operating in heavy traffic.

A simple probabilistic interpretation of the conditional expectation of the queue length in terms of $z^\star$ is possible. If the queue length distribution is assumed to exhibit a geometric tail behavior following L, i.e, $\Pr(q > L) = \alpha$ and

$$\Pr(q > i + L) = \alpha\beta^i, \quad i \geq 0 \tag{15}$$

so that

$$E[q - L \mid q > L] = \frac{\beta}{1-\beta} = \frac{1}{z^\star - 1} \tag{16}$$

Determination of Dominant Root $z^\star$

Let the NXN matrix $\mathbf{A}(z)$ denote the p.g.m of the arrival process. Denote $\lambda_i(z)$, $1 \leq i \leq N$, as the N eigenvalues of $\mathbf{A}(z)$. Then we have the following factorization of the determinant:

$$\Delta(z) = \mid z\mathbf{I} - \mathbf{A}(z) \mid = \prod_{i=1}^{N}(z - \lambda_i(z)) \tag{17}$$

Therefore, once the eigenvalues of the p.g.m of the source are given, the zeros of the determinant $\Delta(z)$ are easily obtained. Now it remains to determine which of the N equations $z = \lambda_i(z)$ solves $z^\star$. It can be shown that the root $z^\star$ solves the equation $z = \mathcal{X}(z)$ where $\mathcal{X}(z)$ is the Perron-Frobenious (PF) eigenvalue of $\mathbf{A}(z)$. This eigenvalue is simple, has the largest modulu, is an analytic function of z, and has many other interesting properties including that $\mathcal{X}(1) = 1$, and $\mathcal{X}'(1) = \rho$ [8] (as pointed out ρ denotes the average number of arrivals in a slot (hence the utilization of the single server queue.)

For an arrival process which is the superposition of independent arrival processes, the eigenvalues of the superposition process is easily obtained in terms of those of the individual processes. More importantly, the PF eigenvalue of the superposition process is simply the *product* of the PF eigenvalues if the individual sources.

In this paper, we concentrate on the important special case which a source is either "OFF" or "ON". When the source is OFF it does not transmit any cell, and when source is ON it transmits at a constant rate one cell per slot. We assume that the duration of the OFF period (t_{off}) is geometrically distributed with mean $E[t_{off}]$ cells, and the ON period has a "mixed-geometric" (mixture of two geometric) distribution. The parameters of this distribution are chosen to fit the first two moments of t_{on}.

Based on the discussion above, it is easy to see that such a source can be characterized by three states 0, 1 and 2 with the following probability generating matrix

$$\mathbf{A}(z) = \begin{bmatrix} p_{00} & p_{01}z & p_{02}z \\ p_{10} & p_{11}z & 0 \\ p_{20} & 0 & p_{22}z \end{bmatrix} \tag{18}$$

In the OFF state (state 0) the source does not transmit any cell, and in the ON state (states 1 or 2) it transmits one cell.

Based on the above description of the source, the probability generating function of the ON and OFF periods are given by

$$E[z^{t_{off}}] = \frac{(1-p_{00})z}{1-p_{00}z} \tag{19}$$

$$E[z^{t_{on}}] = \frac{p_{01}}{p_{01}+p_{02}} \frac{(1-p_{11})z}{1-p_{11}z} + \frac{p_{02}}{p_{01}+p_{02}} \frac{(1-p_{22})z}{1-p_{22}z} \tag{20}$$

Now it remains to characterize the source parameters based on its characteristics. We have four independent parameters p_{01}, p_{10}, p_{02} and p_{20}. Assuming that the mixed-geometric distribution has a balanced mean, i.e, the mean times that the source stays in states 1 and 2 are equal, we have

$$\frac{p_{01}}{p_{10}} = \frac{p_{02}}{p_{20}} \tag{21}$$

To find the remaining three independent parameters, we use ρ to denote the utilization of the source, T for average ON period, and finally C^2 the squared coefficient of variation (variance divided by mean squared). It can be shown that for the source characterized by the triplet (ρ, T, C^2), we have

$$p_{10},\ p_{20} = \frac{1}{T}\left[1 \pm \sqrt{1 - \frac{2T}{(1+C^2)\,T+1}}\right],\quad C^2 \geq 1 - 1/T \tag{22}$$

$$p_{01} = \frac{\rho}{2(1-\rho)} p_{10},\quad p_{02} = \frac{\rho}{2(1-\rho)} p_{20} \tag{23}$$

and the obvious identities

$$p_{00} = 1 - p_{01} - p_{02},\quad p_{11} = 1 - p_{10},\quad p_{22} = 1 - p_{20} \tag{24}$$

The flexibility of the mixed-geometric burst size distribution is quite evident. It allows to model sources with arbitrary first two moments. In the special case where

$C^2 = 1 - 1/T$, the mixed-geometric source described above is equivalent to a *Binary Markov Source which has the restriction of having a squared coefficient of variation of less than one for the burst size duration.*

Now consider the superposition of K sources where the ith source is characterized by the triplet (ρ_i, T_i, C_i^2). Denote the total utilization by ρ defined by $\sum_{i=1}^K \rho_i$. Before we provide the main result on the behavior of the dominating root z^*, we introduce some notations. Let $T_i = \nu_i T$ and $\mathcal{X}_i^{(n)}(1;T)$ denote the n th derivative of the PF eigenvalue of source i (as function of T) and

$$\gamma_i^{(n)} \stackrel{\Delta}{=} \lim_{T\to\infty} \frac{\mathcal{X}_i^{(n)}(1;T)}{T} \tag{25}$$

It can be shown that for the Ternary Markov Source, $\gamma_i^{(n)}$ exists for all n and it is finite. Using the fact that the PF eigenvalue of $\mathbf{A}(z)$ solves the polynomial equation (in $\mathcal{X}(z)$) generated from the determinant

$$|\mathbf{A}(z) - \mathcal{X}(z)\mathbf{I}| = 0 \tag{26}$$

for $n = 2$ and $n = 3$ we can show

$$\gamma_i^{(2)} = \rho_i(1-\rho_i)^2(1+C_i^2)\nu_i \tag{27}$$

$$\gamma_i^{(3)} = 3\rho_i(1-\rho_i)^3(1+C_i^2)[(1-\rho_i)C_i^2 - \rho_i]\nu_i^2 \tag{28}$$

It is possible to show the following: Let T_i the average ON period of source i be equal to $\nu_i T$. For $\rho < 1$, z^* allows the following Laurent expansion in terms of T

$$z^*(T) = 1 + c_1\left(\frac{1}{T}\right) + O\left(\frac{1}{T^2}\right) \tag{29}$$

$$c_1 = \frac{2(1-\rho)}{\sum_{i=1}^K \gamma_i^{(2)}} + O(1-\rho)^2 \tag{30}$$

The above theorem suggests the following simple approximation in heavy traffic (by substituting $\nu_i = T_i/T$)

$$z^*_{approx1} \approx 1 + \frac{2(1-\rho)}{\sum_{i=1}^K \rho_i(1-\rho_i)^2(1+C_i^2)T_i} \tag{31}$$

The expansion for z^* given in the above theorem can be enhanced by taking more terms. For example we can show that

$$c_1 = \frac{2(1-\rho)}{\sum_{i=1}^K \gamma_i^{(2)}} - \frac{4(\sum_{i=1}^K \gamma_i^{(3)})\,(1-\rho)^2}{3(\sum_{i=1}^K \gamma_i^{(2)})^3} + O(1-\rho)^3 \tag{32}$$

which suggests the following approximation for the dominating root z^*

$$\begin{aligned} z^*_{approx2} \approx\ & 1 + \frac{2(1-\rho)}{\sum_{i=1}^K \rho_i(1-\rho_i)^2(1+C_i^2)T_i} \\ & - \frac{4\sum_{i=1}^K \rho_i(1-\rho_i)^3(1+C_i^2)[(1-\rho_i)C_i^2 - \rho_i]T_i^2\,(1-\rho)^2}{[\sum_{i=1}^K \rho_i(1-\rho_i)^2(1+C_i^2)T_i]^3} \end{aligned} \tag{33}$$

It should be noted that in the heavy traffic, as ρ approaches one, most of the contribution is in the first two terms of the above expansion.

Note that in both (31) and (33) the impact of each source and their characteristics, i.e., utilization, mean and variance (or equivalently the squared coefficient of variation) of the burst period of each source is very clear and appears in a simple manner.

Also Note that the above result is surprisingly simple and it captures the different characteristics of each source on the dominating root very clearly. In the special case of Binary Markov Sources, i.e., $C_i^2 = 1 - \frac{1}{T_i} \approx 1, \forall i$ (large T_i is assumed), we recover the result in [11].

In the model which has been discussed so far, it has been assumed that the source can transmit at the peak rate of one cell per slot. In the next section we relax this assumption.

Slow Sources

Now consider the same source, but with peak rate which is less than one cell per slot. For source i, $1 \leq i \leq K$, the source is constrained to generate a maximum of one cell every R_i slots. However, in every R_i slots, it behaves exactly the same as the Ternary Markov Source (TMS) described above, i.e., the TMS source (with $R_i = 1$) is "imbedded" in every R_i slots. Obviously the utilization of this source is $\frac{\rho_i}{R_i}$. Therefore, the cell arrival process in our model is characterized by the quadruplet $(\rho_i, R_i, T_i, C_i^2)$. This model provides a uniform framework for dealing with *individual* discrete sources of different peak rates which are slower than the server (link) rate. (For example, if the link rate is 150 Mb/s, and the peak rate of a source is 10 Mb/s, then the source can transmit a maximum of one cell in every 15 slots, so that for this source R_i is simply 15.) It should be note that based on above description that when $\rho_i = 1$ corresponds to a periodic source. In this case, when all the sources are identical, exact analysis has been reported by number of researchers, e.g., [3].

It should noted that the underlying assumption in our analysis is that the queue length distribution has an infinite support. This is certainly true in the case where all the peak rates are one cell per slot and there are at least two sources feeding the multiplexer. However, in the case where the slow sources are considered, this may only be true if the sum of the peak rates of individual sources will exceed the link peak rate. Since it is also assumed that the queueing process is stable, the following two conditions will have to satisfy

$$\sum_{i=1}^{K} \frac{1}{R_i} > 1 \tag{34}$$

$$\rho \triangleq \sum_{i=1}^{K} \frac{\rho_i}{R_i} < 1 \tag{35}$$

It can be shown that the PF eigenvalue of the slow source described above is equal to $\mathcal{X}_i^{\frac{1}{R_i}}(z)$ where $\mathcal{X}_i(z)$ is the PF eigenvalue of the original fast source (i.e., source with $R_i = 1$). Now if the queueing system is stable and its distribution has infinite support, we can show the corresponding result for the dominating root (taking only two terms) is given by

$$z^* \approx 1 + \frac{2(1-\rho)}{\sum_{i=1}^{K} \frac{\rho_i}{R_i}(1-\rho_i)^2(1+C_i^2)T_i} \tag{36}$$

Applications

In what follows, we discuss the application of our results in the heavy traffic based on the approximation in (36). Similar to the fast sources, enhanced approximations for the slow sources can be obtained easily.

An immediate application of our results is in buffer sizing of a heterogeneous multiplexer fed by very bursty arrivals. Utilizing the simple approximation given by (36) (or enhanced approximations) various approximations for the queue length distribution is possible depending on the range and extent that a geometric tail behavior is assumed. Since the $\Pr(q > 0) = \rho$, ($\rho = \sum_{i=1}^{K} \frac{\rho_i}{R_i}$), the simplest approximation on the queue length distribution would be the one assuming that the entire distribution following $i = 0$ is geometric with parameter $\beta = \frac{1}{z^*}$, i.e.,

$$\Pr(q > i) \approx \rho\beta^i \approx \rho \left[1 + \frac{2(1-\rho)}{\sum_{i=1}^{K} \frac{\rho_i}{R_i}(1-\rho_i)^2(1+C_i^2)T_i}\right]^{-i} \tag{37}$$

Some discussion is in order here. First, as expected, this result is most accurate when the multiplexer is operating in the heavy traffic. This Second, it is based on infinite (large) buffer assumption which can easily be shown to *overestimate* the actual buffer occupancy. But, if the buffer is large and small probability of loss is desired (e.g., ATM networks), it is usually a good approximation to the finite buffer case.

The second and the main application of our simple result is in call admission control in high-speed networks, e.g., ATM. Modeling an high-speed link as a heterogeneous multiplexer is common [5, 6, 7]. As discussed, it is desirable to define a simple and effective metric which takes into account a connection characteristic and based on its required grade-of-service, it indicates if that connection can be accepted. In this setting, for a given buffer size B (in terms of number of cells), a connection is accepted if the resulting loss probability (in our heterogeneous multiplexer) is less than a desired small number say ϵ. Using our approximation for the multiplexer cell occupancy, we get the following metric for the call admission control. The controller does NOT accept a connection unless the following is true:

$$Ln(\rho) - BLn(\left[1 + \frac{2(1-\rho)}{\sum_{i=1} \frac{\rho_i}{R_i}(1-\rho_i)^2(1+C_i^2)T_i}\right]) < Ln(\epsilon) \tag{38}$$

where Ln denotes the natural logarithm.

Somewhat even simpler metric for the call admission control policy may be given if we approximate the above result assuming that $\rho \approx 1$. We get the following simple result ($\kappa \stackrel{\Delta}{=} - Ln(\epsilon)$)

$$(\frac{\kappa}{1-\rho} - 1)\sum_{i=1} \frac{\rho_i}{R_i}(1-\rho_i)^2(1+C_i^2)T_i < 2B \tag{39}$$

Therefore, in either (38) or (39) we only have to update only two parameters, namely, the total utilization $\rho \stackrel{\Delta}{=} \sum_{i=1} \rho_i$, and the quantity $\sum_{i=1} \frac{\rho_i}{R_i}(1-\rho_i)^2(1+C_i^2)T_i$ each time a connection is taken down. To see if a new connection can be accepted, the two parameters are calculated to see if the above inequality will still hold, if it does, the connection is accepted, and the two parameters are accordingly updated.

Numerical Results and Discussions

In this section we first evaluate the accuracy of the two approximations (31) and (33). We assume two sets of sources where set i consists of K_i identical sources, $1 \leq i \leq 2$.

The sources within set i are identified by the triplet (ρ_i, T_i, C_i^2). $R_i = 1$ is assumed unless otherwise stated. It is also assumed that $T_i = \nu_i T$, where results for different values of T are given.

We provide the result on $1/(z^* - 1)$. The exact computation of z^* was carried out by the exact solution of $\mathcal{X}(z)$ (which solves a cubic polynomial) followed by iterative solution of the equation $z = \mathcal{X}(z)$. We obtain the following numerical results for the symmetric case $K_1 = K_2 = 10$ and $(\rho_i, T_i, C_i^2) = (.045, T, 2.)$, $i = 1, 2$.

T	Exact	Approx. 1	Approx. 2
2	28.23	24.62	28.78
10	141.14	123.12	143.95
18	254.08	221.62	259.10

The corresponding result for $r_i = .040$ (total utilization of .8):

T	Exact	Approx. 1	Approx. 2
2	14.88	11.06	16.42
10	73.94	55.30	82.09
18	133.02	99.53	147.76

To examine a few heterogeneous environments, we assume $K_1 = 10$, $K_2 = 20$, $(\rho_1, T_1, C_1^2) = (.035, T, 2)$ and $(\rho_2, T_2, C_2^2) = (.0275, T, 3)$.

T	Exact	Approx. 1	Approx. 2
2	35.78	30.58	36.55
10	178.48	152.92	182.76
18	321.20	275.26	328.97

For the second heterogeneous environment we examine we let $K_1 = 10$, $K_2 = 20$, $(\rho_1, T_1, C_1^2) = (.040, T, 2)$ and $(\rho_2, T_2, C_2^2) = (.020, T, 3)$. We get

T	Exact	Approx. 1	Approx. 2
2	18.44	13.21	20.85
10	91.58	66.06	104.24
18	164.72	118.92	187.63

Finally for the last case we assume $K_1 = 10$, $K_2 = 20$, $(\rho_1, T_1, C_1^2) = (.045, T, 2)$ and $(\rho_2, T_2, C_2^2) = (.0225, 2T, 3)$. We get

T	Exact	Approx. 1	Approx. 2
2	56.55	46.71	63.10
10	282.37	233.55	315.52
18	508.21	420.40	567.94

The above tables show that as expected, the approximations are most accurate in the heavy traffic. They also clearly indicate that $1/(z^* - 1)$ is approximately linear in T. This means that most of the contribution of of the expansion $z^* - 1 = \frac{c_1}{T} + O(\frac{1}{T})$ is indeed in c_1. Therefore, the approximations for z^* can be furthere enhanced by taking even more terms in (32) which involves the calculation of higher order (larger n) $\gamma_i^{(n)}$. Our numerical results show that the same conclusion may be drawn for slow sources. However, for this sources, a geometric tail behavior usually exhibits in higher percentiles than fast sources.

Conclusions

In this paper a mathematical model for sources with highly-bursty traffic was discussed and analyzed. We derived a simple heavy traffic approximation for the queue length of a multiplexer fed by non-identical such sources. Our result provides a simple and effective algorithm for admission control which can be readily implemented in real time. The algorithm uses a metric which uses only the utilization and the first two moments (average and square coefficient of variation) of the burst period of such sources.

References

[1] D. Anick, D. Mitra, and M.M. Sondhi,"Stochastic Theory of a Data Handling System with Multiple Sources," *The Bell System Technical Journal,* Vol. 61, No. 8, Oct. 1982, pp. 1871-94.

[2] K. Bala, I. Cidon and K. Sohraby, "Congestion Control for High-Speed Packet Switch," *INFOCOM'90*, June 1990.

[3] A. Bhargava, P. Humblet, and M.G. Hluchyj, "Queueing Analysis of Continuous Bit-Stream Transport in Packet Networks," *GLOBCOM' 89.*

[4] J. N. Daigle, Y. Lee and M. N. Magalhaes, "Discrete Time Queues with Phase Dependent Arrivals," *INFOCOM'90*, pp. 728-732, June 1990.

[5] R. Guerin, H. Ahmadi and M. Naghshineh, "Equivalent Capacity and Its Application to Bandwidth Allocation In High-Speed Networks," *IEEE JSAC*, Vol. 9, No. 7. Sep. 1991.

[6] Y. Miyao,"A Call Admission Control Scheme in ATM Networks," *ICC'91.*

[7] T. Murase, H. Suzuki, and T. Takeuchi,"A Call Admission Control For ATM Networks Based on Individual Multiplexed Traffic Characteristics." *ICC'91.*

[8] M. F. Neuts, *Structured Stochastic Matrices of the M/G/1 Type and Their Applications*, New York, Marcel Dekker Inc., 1989.

[9] M. F. Neuts and Y. Takahashi, "Asymptotic Behavior of the Stationary Distributions in the $GI/PH/c$ Queue with Heterogeneous Servers," *Wahrscheinlichkeitstheorie*, Vol. 57, pp. 441-452, 1981.

[10] I. Stavrakakis,"Analysis of a Statistical Multiplexer under a General Input Traffic Model," *INFOCOM'90.*

[11] K. Sohraby,"On the Asymptotic Behavior of Heterogeneous Statistical Multiplexer with Applications," *INFOCOM'92.*

[12] K. Sohraby and M. Sidi, "On the Performance of Bursty and Correlated Sources Subject to Leaky Bucket Rate-Based Access Control Schemes," *INFOCOM '91.*

[13] G. M. Woodruff and R. Kositpaibbon,"Multimedia Traffic Management Principles for Guaranteed ATM Network Performance," *IEEE Trans. Select. Areas in Communications*, Vol. JSAC. 8, Apr. 1990.

A USER RELIEF APPROACH TO CONGESTION CONTROL IN ATM NETWORKS

Ioannis Stavrakakis, Mohamed Abdelaziz, and David Hoag

Department of Electrical Engineering and
Computer Science
University of Vermont
Burlington, Vermont 05405

ABSTRACT

In this paper a new, insightful description of the object functions of a Traffic Regulator (TR) - implementing a model traffic behavior (delivered to the network) and a (user) relief function - is introduced. By identifying the relief function associated with the Leaky Bucket (LB), an explanation for the inefficiency of this widely studied TR is presented. The new class of $\rho - relief\ LB\ TRs$ is then developed by focusing on the implementation of more flexible and efficient relief mechanisms. The improved effectiveness of the proposed class is established through a comprehensive comparison with the standard LB. The results from this study, as well as the slotted nature of the ATM environment, have motivated the introduction and study of the $\sigma - relief\ TR$ implementing a near-periodic spacer. The induced cell loss probabilities have been derived through the analysis of a finite capacity queueing system with service opportunities determined in terms of a periodic pattern consisted of $C + 1$ subframes.

INTRODUCTION

Preventive control is widely considered to be the most promising approach for traffic congestion management in the emerging high-speed Asynchronous Transfer Mode (ATM) networks. An extended survey of mechanisms - called bandwidth enforcement mechanisms or traffic regulators (TR) - developed for the implementation of such control may be found in the references[1]. The object function of a TR is to control the flow of the user traffic to the network in a way that unacceptable network congestion be avoided; no network state information is assumed to be available to the users.

The prevailing approach to the design of efficient TRs has been the following: At first, certain key measures of basic traffic characteristics - which have a significant impact on the network performance - are identified. For instance, such measures can be the cell rate and burst length of the user traffic. Then, the user and the network agree on some limits on the values of these measures (contract), so that the network be able to plan based on the maximum expected amount of stress (potential for congestion) coming from the particular

Asynchronous Transfer Mode Networks, Edited by Y. Viniotis
and R.O. Onvural, Plenum Press, New York, 1993

user. Network congestion may then be controlled through an effective TR, whose basic function is to provide for smoothing of the user traffic and prevention of network congestion. Smoothing of the traffic occurs by not allowing the values of key measures associated with the actual traffic delivered to the network to exceed some predetermined limits. A TR may also be seen as a mechanism which enforces compliance with the contract agreements.

The most widely studied and promising TR is the Leaky Bucket (LB)[1-15]. Other window-type TRs - such as the jumping, triggered jumping and the moving windows, as well as the exponentially weighted moving average - have also been proposed without significant (if any) advantages over the LB[5]. The inefficiency of the LB at both the user and network premises has been established in several studies[8-11]. Modifications of the classical LB have been proposed to cope with these inefficiencies, as well as its usage in conjunction with other schemes[16-21].

The $\rho - relief$ LB studied in this paper may be seen as a leaky bucket equipped with a spacer. The potential increase of the LB efficiency, through the introduction of a spacer, has been speculated[17,18]. The inefficiency of a LB (without token buffering capability) in preventing cell clusters from entering the network has been reported and the impact of a worst case scenario has been illustrated[20]. Based on the latter, it is suggested that a completely different approach be followed. The spacer-controller is proposed as a means of enforcing the peak rate of the traffic entering the network; two spacing algorithms were presented. In this work, a totally different philosophy is followed, leading to the implementation of the proposed relief-based class of traffic regulators. Their efficiency is based on the efficiency of the implemented relief mechanism and it is illustated though numerical results.

The outline of the paper is as follows. In the next section the LB mechanism is briefly described and its efficiency problems are discussed. A new, insightful description of the basic functions of a TR is presented and the LB's problems are explained in terms of the inefficiency of the implementation of these functions. In the sequel, the $\rho - relief$ LB is presented and its potential for efficiency is both discussed and illustrated through a simulation study. The conclusions from the simulation study have motivated the consideration of the σ- relief mechanism. This mechanism is similar to the space-controller[20]. The cell loss probabilities induced by this mechanism are analytically derived through the formulation and analysis of a general queueing system. Numerical results further illustrate the effectiveness of the relief-based mechanisms. Finally, some concluding remarks are presented in the last section.

THE LEAKY BUCKET AND ITS (INHERENT) EFFICIENCY PROBLEMS

A queueing model of the (generalized)[4] LB is shown in Fig. 1. The cell traffic delivered to the network is controlled by means of tokens. Tokens arrive to the token pool (of capacity b, $0 \leq b \leq \infty$) periodically with period T (slots). Cells are temporarily stored in the Cell Queue (CQ) (of capacity c, $0 \leq c \leq \infty$). A cell (token) that finds the queue (token pool) full is discarded. In order for a cell to be delivered to the network, it must obtain a token from the token pool. If the token pool is empty, the cell must wait in the queue until a token enters the token pool. In the first version of the LB, the queue is absent $(c = 0)$. In this case, cells are discarded if upon arrival the token pool is empty and the cell loss probability is maximized. By allowing for some cell queueing $(c > 0)$, the cell loss probability is reduced at the expense of the introduction of cell waiting time.

It is easy to establish that the maximum cell rate (base-rate) of the traffic delivered to the network is equal to $R = \frac{1}{T}$ and its maximum burst length $b' = b + 1$ for $T > b + 1$ (slightly larger if $T \leq b + 1$). A sufficiently long period of inactivity allows for the accumulation of b tokens in the token pool. A sufficiently large burst of cell arrivals to a full token pool will

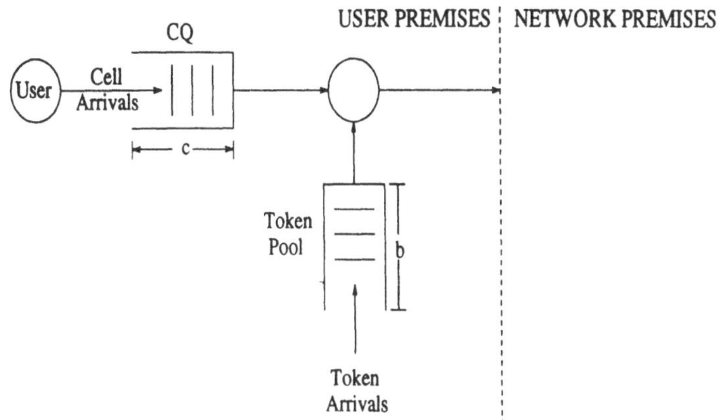

Figure 1: A queueing model for the Leaky Bucket.

result in a burst of maximum length b' in the cell traffic delivered to the network. Thus, the LB is a traffic regulator which enforces the rate and burst length of the cell traffic delivered to the network.

The LB mechanism has been studied extensively under both uncorrelated and correlated user traffic arrival processes and for any capacity of the cell queue, ranging from zero to infinity[3-15]. It has been clearly concluded - especially under correlated user traffic - that the LB is very effective in preventing network congestion when the base-rate R is set very close to the mean rate of the user traffic and the maximum burst length is kept relatively small. In this case, though, the induced cell loss and/or delay at the user premises is, in most cases, unacceptable. Acceptable cell loss and/or delay at the user premises may be achieved by setting the base-rate R close to the peak rate of the user traffic and/or setting the token pool capacity to a very large value. Under such settings, tokens will be almost always available and the LB becomes virtually ineffective (no significant traffic smoothing can be achieved).

In view of the above, it seems that there is an inherent inefficiency of the LB associated with its basic functions (enforcing rate and burst length). Some insight into the problems associated with the basic functions of the LB is presented in the sequel, in terms of the effectiveness of these functions in implementing the object functions of a TR as defined below.

By refocusing on the primary objective in designing a TR - that is, the prevention of network congestion while maintaining the Quality of Service (QoS) of the regulated users - two object functions associated with the operation of a TR may be easily defined:

(a) A desired traffic <u>model behavior</u> (for the user traffic delivered to the network) should be identified, so that the <u>network stress</u> (potential for network congestion) be minimized. In this case, the amount of QoS deterioration at the network premises will be minimized. The <u>primary function</u> of a TR should then be to implement (enforce) this model behavior.

(b) It is expected that the enforcement of a rigid, network-centric model behavior - through the primary function of a TR - will induce significant <u>user stress</u> (potential for QoS deterioration at the user premises), which may result in unacceptable QoS. To reduce the significant user stress due to the primary function of a TR, a (user) <u>relief mechanism</u> should be adopted. The relief to the user will be provided through a controlled allowance for divergence of the traffic delivered to the network from the model behavior. The implementation of an appropriate relief mechanism should be the <u>secondary function</u> of a TR.

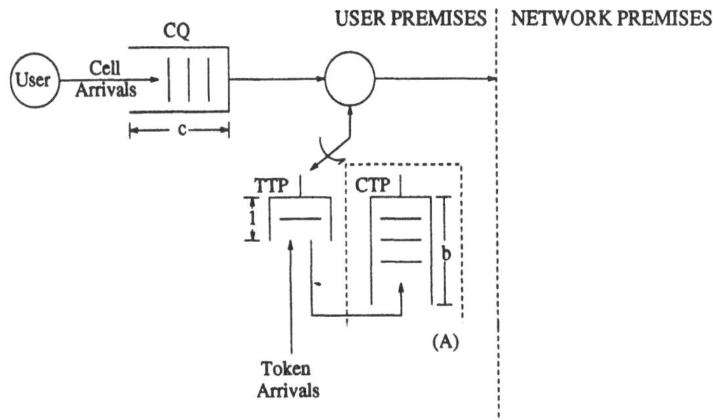

Figure 2: A queueing model for the Leaky Bucket and the $\rho - relief\ LB$.

The traffic model behavior of a TR can be determined by considering an active user and deactivating the relief mechanism or, equivalently, assuming that the intense activity of the user has exhausted the amount of relief which could become available. The user relief mechanism is defined to be the mechanism responsible for the deviation of the traffic delivered to the network from the model behavior.

In the sequel, the inefficiency of the LB is explained in terms of the inefficiency of its secondary function. Throughout this paper time is measured in slots (defined to be equal to the cell transmission time); time k will denote the end of the kth slot. Events, such as cell arrival/departure and token arrival/release, will be defined to occur at the end of the slots.

An equivalent model for the LB mechanism with token pool capacity equal to b is shown in Fig. 2. As it will be explained later, the LB considered here (Fig. 2) is slightly different from the one considered in the past (Fig. 1) with the same capacity (b). It will be easily seen that the LB of Fig. 2 may provide more tokens than that of Fig. 1 but less than that of Fig. 1 with token pool capacity b+1. This difference between the two models is insignificant, though. To facilitate the description of the primary and secondary functions associated with the operation of the LB, as well as those of the proposed $\rho - relief\ LB$, the model shown in Fig. 2 will be adopted throughout this paper. The operation of the LB mechanism (Fig. 2) is completely determined in terms of the token arrival and token release protocols; a token release enables the transmission of the cell at the head of CQ.

Token Arrival Protocol: A token arrives to the Temporary Token Pool (TTP), where it may be kept for at most one slot. Then, if not released, it is forwarded to the Credit Token Pool (CTP) before the next time instant (slot). If CTP is full, the token is discarded. The capacities of TTP and CTP are equal to 1 and b, respectively.

Token Release Protocol: Let $[x]_k$ denote the content of entity x at time k. Let t be the time when a cell is forwarded to the head of CQ.

(a) If $[\mathrm{TTP}]_t = 1$, the token is released from TTP immediately (and, thus, the cell at the head of CQ is delivered to the network).

(b) If $[\mathrm{TTP}]_t = 0$ and $[\mathrm{CTP}]_t = 0$, the token is released from TTP as soon as it arrives.

(c) If $[\mathrm{TTP}]_t = 0$ and $[\mathrm{CTP}]_t > 0$, a token is released from CTP immediately.

In view of the previous description of the LB (Fig. 2) the associated (traffic) model behavior and (user) relief mechanism are easily determined. The user relief is provided through the unused (while in TTP) tokens which are contained in CTP; b' (usually equal to $b+1$, as explained earlier) is the maximum amount of relief which can be provided when the user is feeding the network faster than what the traffic model behavior allows for. The traffic model behavior is easily seen to be a periodic traffic with period $T = \frac{1}{R}$. This may be established by deactivating the relief mechanism (or, equivalently, assuming zero amount of relief ([CTP] = 0)) and assuming a non-empty CQ. The relief mechanism of the LB is identified with block (A) (Fig. 2). When block (A) is absent, no relief can be provided and the cell traffic delivered to the network will be the model traffic (when CQ is non-empty). The following inefficiencies of the relief mechanism of the LB may be observed.

From the user's perspective: The relief mechanism does not seem to be applied at the times when user's relief is most needed. After a sufficiently long period of inactivity, the full potential of the relief mechanism is applied, since the maximum number of tokens (amount of relief or credit) is accumulated in CTP. At that time there is no backlog at the user's premises and, thus, relief may not be needed. On the other hand, after the initial stage of a period in which the CQ is non-empty and the backlog is building up, the relief mechanism is never applied, since CTP becomes soon empty and no more credit can be available. This is a major inefficiency of the LB, as reported in the simulation studies in a different context[8].

From the network's perspective: The relief mechanism is such that, when activated, it delivers a cell traffic which is very stressful to the network; a block of up to b' cells. If relief is not needed at that time (for instance, immediately after a long period of user inactivity), the network is unnecessarily severely stressed. If relief is needed, there may be some relief mechanism more considerate to the network, still delivering the amount of relief necessary to maintain the QoS.

In view of the above, the relief mechanism of the LB seems to be fairly inconsiderate to both the user and the network. In the next section, a new class of TRs is introduced, based on a more considerate relief mechanism. This class presents increased flexibility regarding the trade-off between user and network stress compared to that of the LB and, as a result, significant potential for improved effectiveness.

THE ρ-RELIEF LEAKY BUCKET

At first the $\rho-relief$ Leaky Bucket ($\rho-relief\ LB$), $0 \leq \rho \leq 1$, is introduced. Then its key characteristics are presented and its potential for improved effectiveness (over the LB) is discussed.

The $\rho - relief\ LB$ is shown in Fig. 2. The token arrival protocol to TTP and CTP is identical to that of the LB, as described in conjunction with its equivalent model presented in the previous section (Fig. 2). The token release protocol of the $\rho - relief\ LB$ is different from that of the LB in the following: The tokens in CTP are released upon demand, provided that at most one such token be released over a time interval of length $\frac{1}{\rho}$. In more detail, the token release protocol of the $\rho - relief\ LB$ is described below. Let t be the time when a cell is forwarded to the head of CQ; let ξ_t denote the time elapsed between time t and the time when the last token before time t was released from CTP.

(a) If $[\text{TTP}]_t = 1$, the token is released from TTP immediately.

(b) If $[\text{TTP}]_t = 0$ and $[\text{CTP}]_t = 0$, the token is released from TTP as soon as it arrives.

(c) If $[\text{TTP}]_t = 0$ and $[\text{CTP}]_t > 0$ then:

(c-1) If $\xi_t \geq \frac{1}{\rho}$, a token is released from CTP immediately.

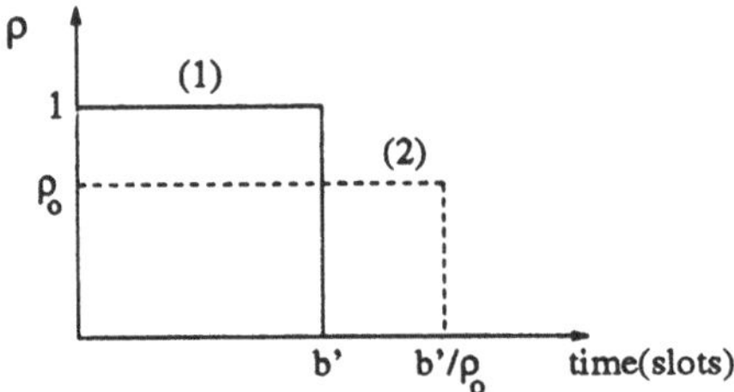

Figure 3: The instantaneous relief functions.

(c-2) If $\xi_t < \frac{1}{\rho}$, a token is released from CTP at time $t + \frac{1}{\rho} - \xi_t$ provided that no token arrives to TTP by that time; otherwise, the token is released from TTP upon arrival.

The traffic model behavior and the relief mechanism of the $\rho - relief\ LB$ may be easily determined. The traffic model behavior is identical to that of the LB and corresponds to a periodic cell traffic delivery to the network when CQ is non-empty. The minimum spacing between consecutive cells delivered to the network according to the traffic model behavior is equal to T. Notice that the network capacity which becomes available to the regulated user is equal to $R = \frac{1}{T}$, when the relief mechanism is deactivated (no credit is available ([CTP] = 0)). The base-rate is also equal to $R = \frac{1}{T}$. Note that the $1 - relief\ LB$ is identical to the LB.

When [CQ]> 0 and [CTP] > 0 the relief mechanism is active. That is, tokens from CTP (credit) become available to the cells upon demand and according to the token release protocol. Let the instantaneous relief be defined as the amount of relief (in tokens) per unit time (slot) provided to the user. When the relief mechanism is inactive, the instantaneous relief is equal to zero. When the relief mechanism is active, it is easy to see that the instantaneous relief of the $\rho - relief\ LB$ is equal to ρ.

The instantaneous relief ρ as a function of time is presented in Fig. 3. Two cases - $\rho = 1$ and $\rho = \rho_0 < 1$ - are shown under the assumption that the same amount of relief b' (in tokens) is provided to the user under both the $1 - relief\ LB$ and the $\rho_0 - relief\ LB$. The duration of this relief interval (given by $\frac{b'}{\rho}$) is equal to b' and $\frac{b'}{\rho_0}$ for the 1-relief and the ρ_0-relief LB, respectively. Notice that there is only one way to provide an amount of relief b' to the user when the LB ($1 - relief\ LB$) is adopted, determined completely by b'. By moving ρ_0 up or down and $\frac{b'}{\rho_0}$ left or right, the same amount of relief b' can be provided to the user in a way which could be less stressful to the network (when ρ decreases, as explained below) and still acceptable to the user.

Let the instantaneous remaining capacity δ associated with a relief mechanism be defined as the unused network capacity (in cells per slot), when the only cell traffic delivered to the network is that provided through the particular relief mechanism; it is equal to one when the relief mechanism is inactive. Clearly, δ may serve as a measure of the instantaneous stress on the network induced by a particular relief mechanism. When $\delta = 0$, the total network capacity is provided to the user operating under the particular relief mechanism. This is a network stress inducing condition, since no capacity can become available to any other network user. Note that $\delta = 1 - \rho$, when the relief mechanism of the $\rho - relief\ LB$ is active. When the relief mechanism of the LB ($1 - relief\ LB$) is active, $\delta = 0$; that is, the relief mechanism of the LB is the most stressful to the network.

Let $C_{LB} = \{\rho - LB_R,\ 0 \leq \rho \leq 1,\ 0 \leq b \leq \infty\}$ denote the class of $\rho - relief\ LB$'s with base rate R (determined by the user cell rate); its elements are determined by setting the instantaneous relief ρ and CTP capacity b. The space of C_{LB} is shown in Fig. 4.

Notice that the sub-class of the LB regulators, $C_{LB-1} = \{1 - LB_R,\ 0 \leq b \leq \infty\}$, corresponds to the boundary of the space at $\rho = 1$. Clearly, the optimal (most efficient)

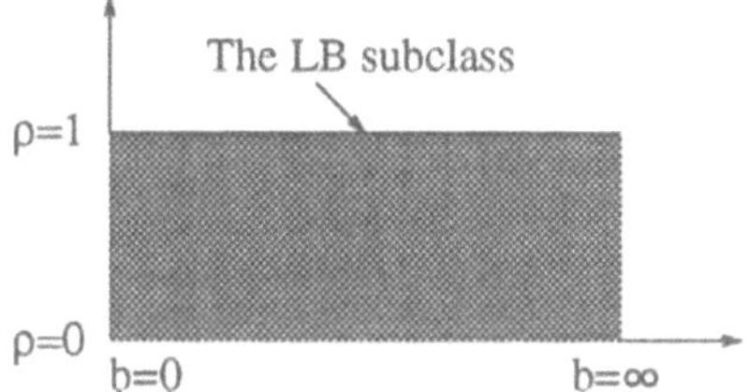

Figure 4: The space of the class $C_{LB} = \{\rho\text{–LB}_R, 0 \leq \rho \leq 1,\ 0 \leq b \leq \infty\}$.

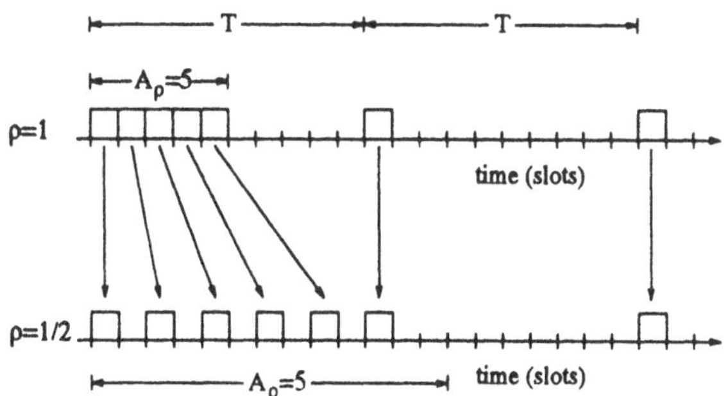

Figure 5: A realization of the cell process delivered to the network.

$\rho - relief\ LB$ (element of C_{LB}) cannot be less efficient than the optimal LB (element of the sub-class C_{LB-1}). As a result, the optimal $\rho - relief\ LB$ for a certain application can only improve the efficiency that the widely considered LB regulator can achieve!

A realization of the cell process delivered to the network when a $1 - relief\ LB$ or a $\frac{1}{2} - relief\ LB$ are in effect, is shown in Fig. 5; the CTP capacity b is assumed to be identical for both regulators. It is easy to establish that a $\rho - relief\ LB$ for $0 \leq \rho < 1$ is more considerate to the network, since the (spread-out) amount of relief A_ρ is provided to the user according to a pattern that is less stressful to the network, compared to that (block-type) associated with the provision of the amount of relief A_ρ in the case of the $1 - relief\ LB$.

From Fig. 5 it is easy to draw the conclusion that a $\rho - relief\ LB$ for $0 \leq \rho < 1$ is less considerate to the user, since relief is provided at a lower rate (compared to that of the LB) and, thus, cells would stay longer in CQ. This conclusion is not necessarily correct. Suppose that the CTP capacity (b_ρ) of a $\rho - relief\ LB, 0 \leq \rho < 1$, is greater than that ($b_1$) of the LB. Then the realization shown in Fig. 6 is possible.

Since $b_\rho > b_1$, A_ρ for $\rho = 1/2$ can be greater than that for $\rho = 1$ and, on the average, cells may stay longer in CQ in the case of the LB, as implied in Fig. 6. Of course if b_1 is

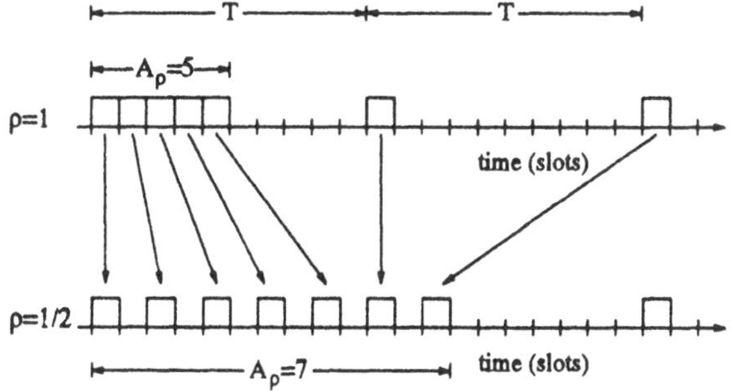

Figure 6: A realization of the cell process delivered to the network.

equal to b_ρ the realization corresponding to that in Fig. 6 would indicate more consideration to the user associated with the $1-relief\ LB$ rather than the $\rho-relief\ LB, 0 \leq \rho < 1$. The key issue, though, is that the network could tolerate a larger duration (by increasing b_ρ) of moderate instantaneous network stress ($\delta > 0$ for $\rho < 1$) compared to the tolerable duration under large instantaneous network stress ($\delta = 0$ for $\rho = 1$).

The above considerations provide for some insight into the operation of the $\rho-relief\ LB$ and its potential for increased efficiency; the latter will be the case when the optimal element in the space shown in Fig. 3 is not contained in the boundary at $\rho = 1$. From the above discussion, it turns out that the desired balance between the user stress and the network stress (to achieve a specified QoS) may be achieved (if at all) by controlling the CTP capacity b_1, when a $1 - relief\ LB$ regulator is adopted. When a ρ-relief LB is in effect, the latter balance may be achieved by controlling the CTP capacity b_ρ and/or the instantaneous relief ρ. As a result, the $\rho - relief\ LB$ is expected to be more efficient than the LB, or it could be successful where the LB fails.

In the next section a simulation study is presented illustrating the behavior of the ρ-relief mechanism and its improved efficiency compared to that of the LB. The induced mean cell delay at both the user and the network premises is the performance measure adopted for the comparison.

A SIMULATION STUDY

The delay results presented in this section will help get insight and show potential for efficiency of the $\rho-relief\ LB$, rather than establish its effectiveness for a specific application. They are derived from computer simulations.

The regulated sources are assumed to be correlated. The generated traffic is assumed to be governed by a discrete-time first-order Markov process. A cell is generated when the Markov chain is in state 1, no cell is generated when in state 0. The burstiness coefficient γ is set equal to .6; $\gamma = p(1,1) - p(0,1)$, where $p(i,j)$ denotes transition probability from state i to state j, $i,j \in \{0,1\}$. The period T of the token arrival process is set equal to 10 slots resulting in a base rate $R = .1$, slightly higher than the source rate $R_s = .09$.

The mean cell delay at the network premises has been calculated assuming the topology shown in Fig. 7; the network buffer is fed by the regulated traffic of 10 symmetric users. The mean cell delay at the user premises, $D_u(\rho, b)$, is presented in Fig. 8 as a function of the

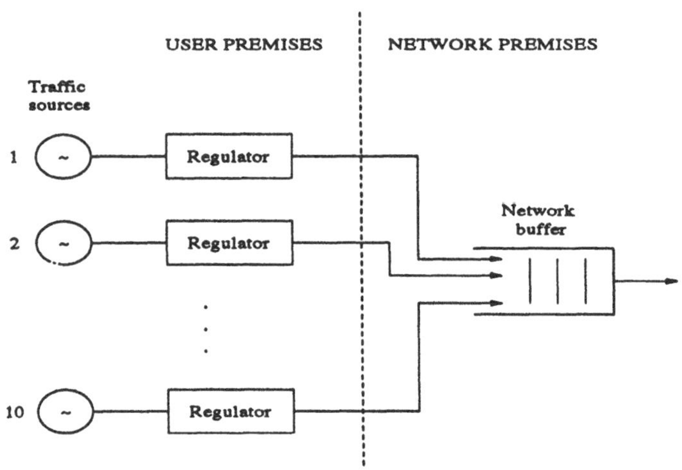

Figure 7: The ATM network link supporting the traffic of 10 regulated users.

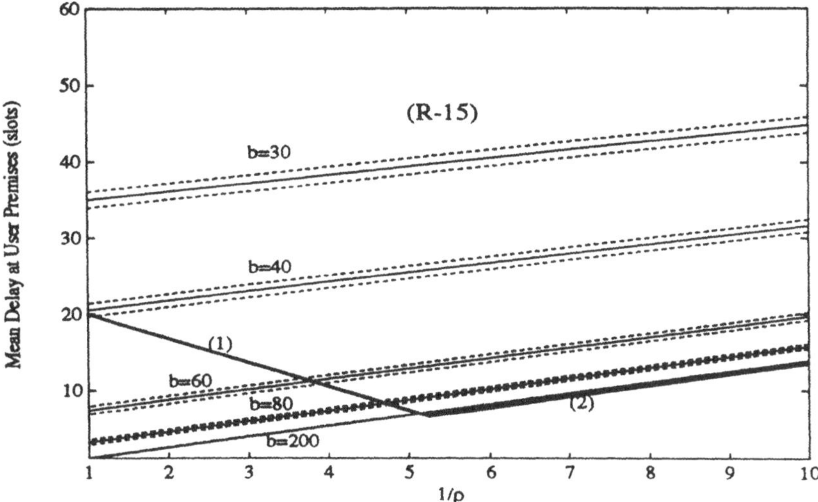

Figure 8: Mean delay results at the user premises ($D_u(\rho, b)$) as a function of the instantaneous relief ρ and for various values of the CTP capacity b; $R = .10,\ R_s = .09,\ \gamma = .60$.

instantaneous relief ρ and for various values of the CTP capacity b. Similar results for the mean delay at the network premises, $D_n(\rho, b)$, are shown in Fig. 9.

Notice that as ρ and/or b increase, $D_u(\rho, b)$ decreases and $D_n(\rho, b)$ increases. The previous behavior is expected since the instantaneous relief and the total amount of relief available to the user increase as ρ and b increase, respectively. Note that the induced stress at the network premises increases as ρ and/or b increase, since the remaining capacity δ decreases and/or the (user) relief duration increases, respectively. As a consequence, the cell delay at the network premises increases.

To illustrate the increased efficiency of the $\rho - relief\ LB$ (for $\rho < 1$) over that of the $1 - relief\ LB$, suppose that QoS considerations impose the constraint that the mean delay induced at the network premises be less than a threshold B_n, say B_n=15 (slots). From Fig.

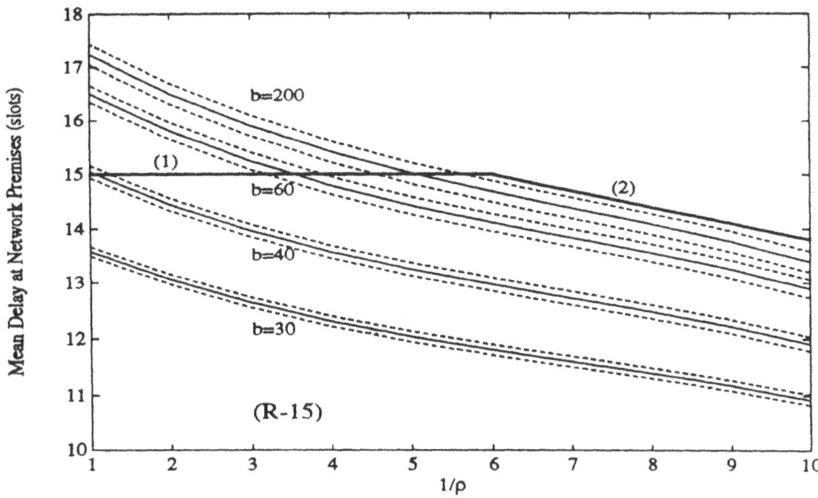

Figure 9: Mean delay results at the network premises ($D_n(\rho, b)$) as a function of the instantaneous relief ρ and for the values of the CTP capacity b; $R = .10,\ R_s = .09,\ \gamma = .60$; N=10 identical users feed the network node.

9, the sub-class $\mathcal{A}_{sym}(B_n)$ of the $\rho - relief\ LB$s which satisfy this constraint (B_n) can be determined; $\mathcal{A}_{sym}(15) = \{\rho - LB_R,\ 0 \leq \rho \leq 1,\ 0 \leq b \leq \infty\ :\ D_n(\rho, b) \leq 15\}$. The operation point of these $\rho - relief\ LBs$ is contained in the region (R-15) (upper bounded by lines (1) and (2) in Fig. 9). The mean delay induced at the user premises when a $\rho - relief\ LB$ from $\mathcal{A}_{sym}(15)$ is adopted, can be determined from Fig. 8, where the acceptable region of operation (R-15) is presented again. Notice that there are $\rho - relief\ LBs$ for $0 \leq \rho < 1$ which induce significantly lower delay than the $1 - relief\ LB$. For instance the $\frac{1}{6} - relief\ LB$ with CTP capacity $b \geq 200$ induces more than 3 times less delay at the user premises than that induced by the $1 - relief\ LB$.

Some comments regarding the optimality of a TR are now in order. As it was stated earlier, the primary objective of a TR is to provide for a mechanism that prevents network congestion. Thus, it is meaningful to define a maximum level of network stress and design TRs which will guarantee it, at least in a probabilistic sense. If the resulting QoS of (some of) the supported users is unacceptable, a smaller collection of such users should be allowed to establish a connection by the call admission policy. The following definition of the optimal element of a class $\mathcal{C}$ of TRs is meaningful, in view of the above discussion. The assumed symmetric environment would render the optimal TR independent from the type of the co-existing users and TRs. The mean delay is considered to be the measure of stress at both the user and network premises.

Definition: Assuming a symmetric environment, the optimal TR in $\mathcal{C}$, β_{sym}, is determined by

$$\beta_{sym} = \arg\{\min_{c \in \mathcal{A}_{sym}(B_n)} \{D_u\}\}$$

where D_u denotes the induced mean cell delay at the user premises and

$$\mathcal{A}_{sym}(B_n) = \{c \in \mathcal{C} : D_n \leq B_n\}.$$

B_n is an upper bound on the mean cell delay at the network premises, D_n. $\mathcal{A}_{sym}(B_n)$ is the collection of TR in $\mathcal{C}$ which deliver to the network a traffic inducing a mean cell delay at the network premises less than B_n.

To establish the potentially increased efficiency of the proposed $\rho - relief\ LB$ $(\rho < 1)$ compared to that of the previously introduced $1 - relief\ LB$, it is important to show that, at least for some cases, the optimal element in $\mathcal{C}_{LB}$ belongs in the sub-class $\bar{\mathcal{C}}_{LB-1} = \{\rho - LB_R,\ 0 \leq \rho < 1,\ 0 \leq b \leq \infty\}$ rather than in $\mathcal{C}_{LB-1} = \{\rho - LB_R,\ \rho = 1,\ 0 \leq b \leq \infty\}$; the latter sub-class contains the standard LBs for various CTP capacities b.

The results presented earlier clearly show that the optimal TR in $\mathcal{C}_{LB}$ is never - for the cases presented there - in $\mathcal{C}_{LB-1}$. $\mathcal{A}_{sym}(15)$ contains all elements in $\mathcal{C}_{LB}$ whose operation point is contained by region (R-15) (Fig. 9). The optimal element is identified by the intersection the boundaries (1) and (2) of the region (R-15). Note that as B_n changes, boundary (1) moves up or down in a parallel fashion in Fig. 9. Similarly, boundary (1) moves up or down in an almost parallel fashion in Fig. 8 as well. As a result, its intersection with boundary (2) (which is part of the delay curve for $b = \infty$) will identify a $\rho - relief\ LB$ which induces lower mean delay at the user premises than that of the optimal $1 - relief\ LB$.

From the simulation study presented in this section, the efficiency of the $\rho - relief\ LB$ with respect to the induced user delay has been illustrated. Another important performance measure of a TR is the induced cell loss probability when finite buffer capacities are available. An analytical study for the derivation of the cell loss probability induced by a practical and potentially better performing version of the $\rho - relief\ LB$ is presented in the next section. Analytical evaluation of cell loss probabilities is of interest due to the difficulty in obtaining simulation results for low loss probabilities.

THE σ - RELIEF TR

From the study presented in the previous section, it appears that the optimal performance of the $\rho - relief\ LB$ is achieved for infinite CTP capacity and a proper value of ρ. Since $\lambda < 1/T$, following the initialization of the system, tokens (credit) will always be available in the CTP. Consequently, the service provided to the CQ through credit tokens can essentially be modeled as periodic with period $1/\rho$. Additional periodic service (with period T) is provided by the tokens from the TTP. The total service rate to the CQ is equal to $1/T + \rho$. By coordinating the two service mechanisms through the replacement of the two periodic service mechanisms by one with period $1/(1/T + \rho)$, it is expected (and was observed) that the performance at both the user and the network premises improves, due to the reduction in the variance of both the service provided to the CQ and the inter-arrival time of the cells delivered to the network. The $\sigma - relief\ TR$ defined below implements the previous ideas.

Consider a TR in a slotted environment which consists of a Cell Queue (CQ) for the temporary storage of the cells generated by the regulated user. CQ is assumed to be served periodically at a rate $\lambda + \sigma$ (cells per slot); λ denotes the cell generation rate; σ is defined to be the additional service rate provided to the user, beyond that identified by its cell generation rate, $0 \leq \sigma \leq 1 - \lambda$; σ will be called the relief rate.

Note that for a given user rate (λ) and relief rate (σ) the service period $1/(\lambda + \sigma)$ can only be approximately implemented in a slotted environment. If $T_1 = \lfloor \frac{1}{\lambda+\sigma} \rfloor$ - where $\lfloor x \rfloor$ denotes the largest integer which is not greater than x - then an appropriate combination of service intervals (subframes) of length T_1 and $T_2 = T_1 + 1$ (in slots) would implement a nearly periodic service pattern delivering a relief rate arbitrarily close to σ, for sufficiently long implementation horizon, while keeping the variance of the length of the service intervals small; service is assumed to be provided over the first slot of each subframe. Let σ_ϵ be defined to be ϵ-close to (the desired) σ in the sense that

$$|\sigma - \sigma_\epsilon| < \epsilon\sigma \tag{1}$$

Let C_1, C_2, $C_1, C_2 \epsilon Z_0^+$, be such that $C_1 + C_2$ is minimum and

$$\left|\sigma - \left(\frac{C_1 + C_2}{C_1T_1 + C_2T_2} - \lambda\right)\right| < \epsilon\sigma \tag{2}$$

Thus a service pattern (horizon) of length $H(C_1, C_2)$ that consists of C_1 subframes of length T_1 and C_2 subframes of length T_2 - in an alternating fashion to the extent that is possible - will provide a relief rate ϵ-close to the selected σ while keeping the number of subframes $C_1 + C_2$ per implementation horizon $H(C_1, C_2)$ to a minimum. The latter is desirable primarily to minimize the numerical complexity, as it will be shown shortly.

In the sequel, a queueing system of potentially wide applicability is analyzed and applied to the study of the $\sigma - relief\ TR$.

Consider a discrete-time, single server, finite capacity queueing system which receives service according to a periodic service pattern (frame) of length T. The periodic frame is assumed to consist of $C + 1$ subframes of length T_u, $0 \leq u \leq C$. The server visits the queue at the beginning of each subframe and provides service to the queue for one slot; then, it switches away from the queue (Fig. 10). The customer (cell) service time is deterministic and equal to one slot. Cell arrivals to the queue are described in terms of the 2-state Markov process considered earlier.

Let $\{I_m, Q_m, R_m\}_{m \geq 0}$ be a process imbedded at the beginning of the subframes. I_m denotes the state of the Markov chain of the cell source at the beginning of the last slot of the $(m-1)\underline{th}$ subframe; let $S = \{0, 1\}$ denote its state space. Cell arrivals due to the visit of the state I_m occur - if any - at the beginning of the last slot of the $(m-1)th$ subframe. Q_m denotes the value of the queue occupancy process at the beginning of $m\underline{th}$

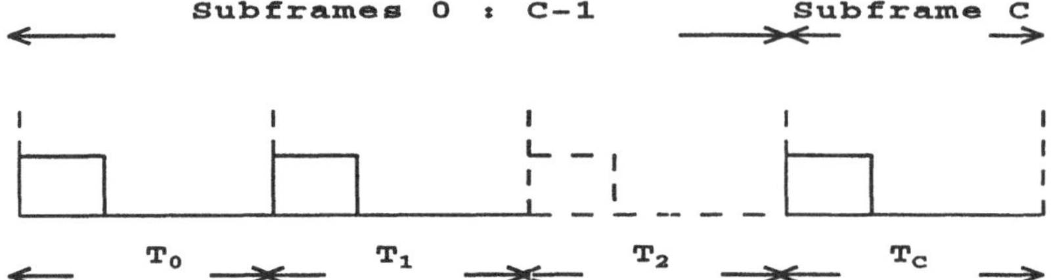

Figure 10: Token Release Pattern

subframe; let $S^Q = \{0, 1, .., Q\}$ denote its state space. R_m represents the index of the $m\underline{th}$ subframe; let $S^R = \{0, 1, ..., C\}$ denote its state space. It can be easily established that $\{(I_m, Q_m, R_m)\}_{m \geq 0}$ is a Markov chain imbedded at the subframe boundaries; its state space is given by $S \times S^Q \times S^R$.

Let $\{M_m\}_{m \geq 0}$ be a sequence of subframe boundaries at which the process $\{(I_m, Q_m, R_m)\}$ visits state $(0, 0, 0)$. Clearly, $\{M_m\}_{m>0}$ is a renewal sequence. Let $\{X_m\}_{m \geq 0}$ denote the sequence of time intervals (in slots) between consecutive renewal points (renewal cycles); let W_m denote the number of cells lost over X_m. Clearly, $\{W_m\}_{m \geq 0}$ is a regenerative process with respect to the renewal process $\{M_m\}_{m \geq 0}$. Let $\overline{X} = E\{X_m\}$ and $\overline{W} = E\{W_m\}$, where $E\{.\}$ denotes the expectation operator. A direct application of the regeneration theorem results in the following expression for the calculation of the cell loss probability[15],

$$L = \frac{\overline{W}}{\lambda \overline{X}} \text{ with probability 1.} \tag{3}$$

where λ denotes the cell arrival rate (per slot) and assuming $\lambda T < 1$ to guarantee that the mean renewal cycle is finite (sufficient condition for a finite capacity queueing system).

In the following, the quantities $\overline{X}$ and $\overline{W}$ are calculated. First, let $X(i, n, u)$ be a random variable denoting the length of time (in slots) between the beginning of the current subframe, at which $\{I_m, Q_m, R_m\}_{m \geq 0}$ is in state (i, n, u), and the next renewal instant from $\{M_m\}_{m \geq 0}$. Let $W(i, n, u)$ be a random variable denoting the number of cells lost over $X(i, n, u)$. The following equations may be derived with respect to $X(i, n, u)$, $(i, n, u) \epsilon S \times S^Q \times S^R$.

For $i \epsilon S, 0 \leq u \leq C - 1$:

$$X(i, 0, u) = \begin{bmatrix} T_u + X(j, k', u+1) & if\ (i \xrightarrow{1} m, a_1 = 0, 1), \\ & (m \xrightarrow{T_u - 1} j, a_{T_u - 1} = k'), 0 \leq k' \leq Q - 1 \\ T_u + X(j, Q, u+1) & if\ (i \xrightarrow{1} m, a_1 = 0, 1), \\ & (m \xrightarrow{T_u - 1} j, a_{T_u - 1} = k'), Q \leq k' \leq T_u - 1 \end{bmatrix} \tag{4}$$

For $i \epsilon S, 0 \leq u \leq C - 1, n \geq 1$:

$$X(i, n, u) = \begin{bmatrix} T_u + X(j, n + k - 1, u + 1) & if\ (i \xrightarrow{T_u} j, a_{T_u} = k), \\ & 0 \leq k \leq Q - n \\ T_u + X(j, Q, u + 1) & if\ (i \xrightarrow{T_u} j, a_{T_u} = k), \\ & Q - n + 1 \leq k \leq T_u \end{bmatrix} \tag{5}$$

For $i \epsilon S$:

$$X(i,0,C) = \begin{bmatrix} T_C & if\,(i \xrightarrow{1} m, a_1 = 0,1), (m \overset{T_C-1}{\rightarrow} 0, a_{T_C-1} = 0) \\ T_C + X(1,0,0) & if\,(i \xrightarrow{1} m, a_1 = 0,1), (m \overset{T_C-1}{\rightarrow} 1, a_{T_C-1} = 0) \\ T_C + X(j,k',0) & if\,(i \xrightarrow{1} m, a_1 = 0,1), (m \overset{T_C-1}{\rightarrow} j, a_{T_C-1} = k'), \\ & 1 \leq k' \leq Q-1 \\ T_C + X(j,Q,0) & if\,(i \xrightarrow{1} m, a_1 = 0,1), (m \overset{T_C-1}{\rightarrow} j, a_{T_C-1} = k'), \\ & Q \leq k' \leq T_C - 1 \end{bmatrix} \tag{6}$$

For $i \epsilon S$:

$$X(i,1,C) = \begin{bmatrix} T_C & if\,(i \xrightarrow{T_C} 0, a_{T_C} = 0) \\ T_C + X(1,0,0) & if\,(i \xrightarrow{T_C} 1, a_{T_C} = 0) \\ T_C + X(j,k,0) & if\,(i \xrightarrow{T_C} j, a_{T_C} = k), 1 \leq k \leq Q-1 \\ T_C + X(j,Q,0) & if\,(i \xrightarrow{T_C} j, a_{T_C} = k), Q \leq k \leq T_C \end{bmatrix} \tag{7}$$

For $i \epsilon S, n \geq 2$:

$$X(i,n,C) = \begin{bmatrix} T_C + X(j,n+k-1,0) & if\,(i \xrightarrow{T_C} j, a_{T_C} = k), \\ & 0 \leq k \leq Q - n \\ T_C + X(j,Q,0) & if\,(i \xrightarrow{T_C} j, a_{T_C} = k), \\ & Q - n + 1 \leq k \leq T_C \end{bmatrix} \tag{8}$$

where $i \xrightarrow{k} j$ denotes a transition of the Markov chain of the source from state i to state j in k steps; a_k denotes the number of cells generated in k consecutive slots. Since the maximum number of cells generated per time slot is equal to one, the maximum number of arrivals over any subframe T_u is equal to T_u, $0 \leq u \leq C$. In Appendix A, some comments on the derivation of the above equations are presented.

By applying the expectation operator to the above system of equations, we obtain the following system of linear equations:

$$\overline{X}(i_1,n_1,r_1) = b(i_1,n_1,r_1) + \sum_{i_2 \epsilon S} \sum_{n_2 \epsilon S^Q} \sum_{r_2 \epsilon S^R} a(i_1,n_1,r_1,i_2,n_2,r_2)\overline{X}(i_2,n_2,r_2) \tag{9}$$

where: $\overline{X}(i,n,r)$ denotes the expected value of $X(i,n,r)$. The linear equations, the coefficients $a(i_1,n_1,r_1,i_2,n_2,r_2)$ and the constants $b(i_1,n_1,r_1)$ are found in Appendix B. Similar equations can be obtained with respect to $W(i_1,n_1,r_1)$. These equations and the coefficients of the resulting system of equations are presented in Appendix C. Since $(0,0,0)$ defines a renewal point, the loss probabilities are found by solving the above systems for $\overline{X}(0,0,0)$, $\overline{W}(0,0,0)$ and using equation (1).

In the sequel, some numerical results illustrating the performance of the $\sigma - relief\ TR$ are presented. The regulated sources and the network topology are assumed to be identical to those considered in the simulation study presented in the previous section. The CQ capacity is assumed to be finite and equal to Q. For the values of σ shown, the maximum number of subfames $C_1 + C_2$ per implementation horizon was 9, for $\epsilon = 0.01$.

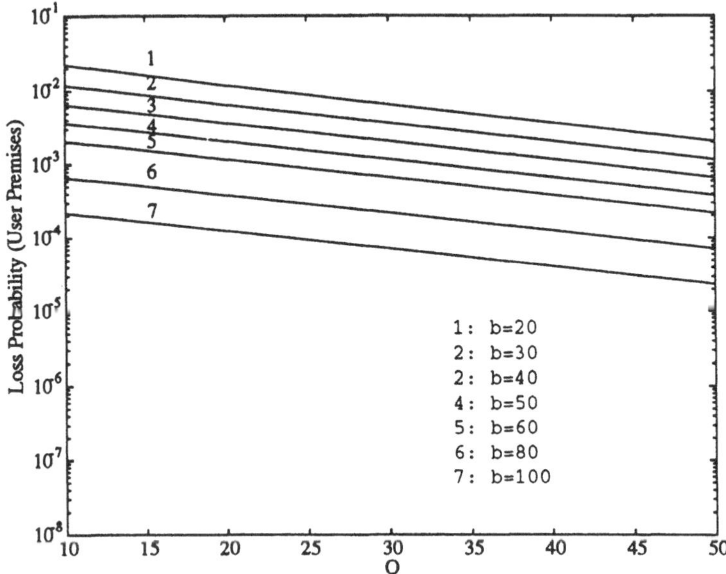

Figure 11: Loss probabilities at the user premises for the $1 - relief\ LB$ for various values of the CTP capacity b.

Fig. 11 and 12 present the cell loss probability at the user premises induced by the $1 - relief\ LB$ and the $\sigma - relief\ TR$, respectively, as a function of Q. Fig. 13 and 14 present the mean delay at the network premises induced by $1 - relief\ LB$ and the $\sigma - relief\ TR$, respectively. Consider a level of performance at the network level achieved by an $1 - relief\ LB$ which provides for some regulation (for instance, for token pool capacity $b < 50$). It is easy to establish that there is a $\sigma - relief\ TR$ which achieves the same or better performance at the network premises by inducing significantly lower loss probability at the user premises, for the same CQ capacity Q.

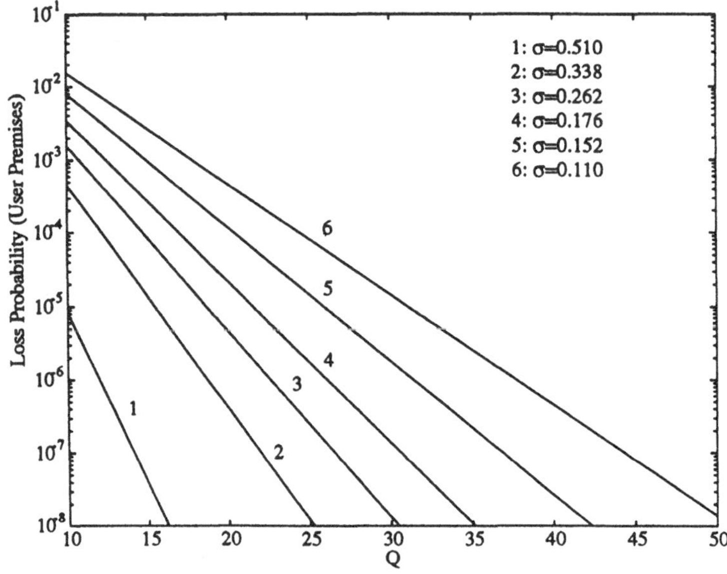

Figure 12: Loss probabilities at the user premises for the $\sigma - relief\ TR$ for various values of the relief rate σ.

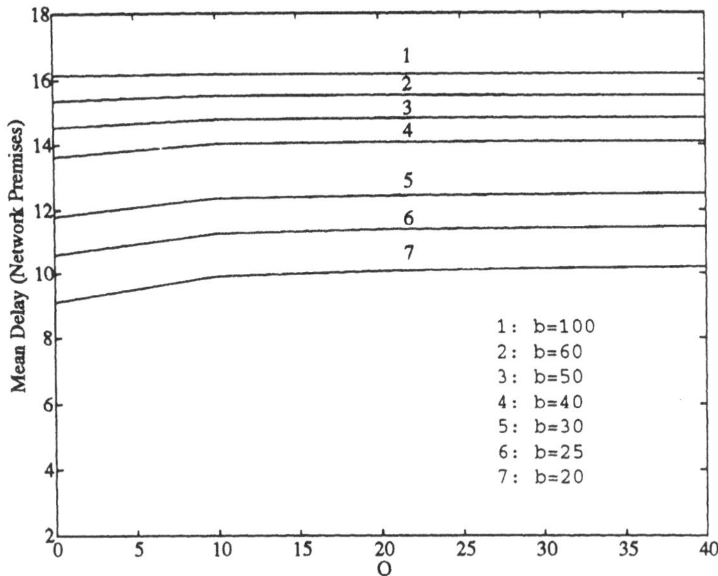

Figure 13: Mean delay results at the network premises for the $1 - relief\ LB$ for various values the CTP capacity b.

Finally it should be noted that departure from the near-periodic spacing by allowing for subframe sizes to differ by more than one time slot has resulted in deterioration of the performance at both the user and the network premises. This has been attributed to the resulting increase of the variance of the service intervals and cell interarrival time to the network queue. Thus, it is concluded that the performance of the window-type policing mechanisms[5] could be improved by spreading uniformly over the window the cells which are permitted to enter the network. By mapping the window to the implementation horizon mentioned above, the performance of window-type policing mechanisms could be evaluated by applying the study of the queueing system considered above.

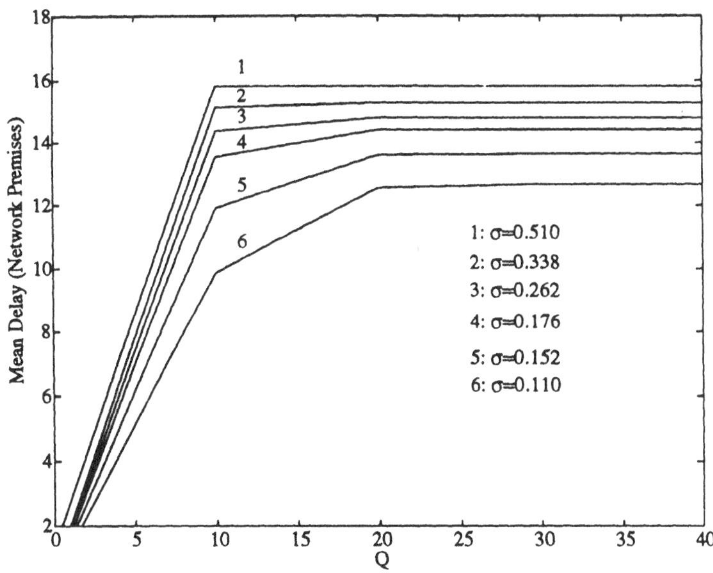

Figure 14: Loss probabilities at the user premises for the $\sigma - relief\ TR$ for various values of the relief rate σ.

SOME CONCLUDING REMARKS

In this paper, a new, insightful description of the object functions of a traffic regulator, implementing a model traffic behavior (delivered to the network) and a (user) relief function, has been introduced. Based on the relief function associated with the leaky bucket, an explanation of the inefficiency of this widely studied traffic regulator was presented. A new class of traffic regulators, the $\rho - relief\ LB$s was introduced by focusing on the implementation of more flexible and efficient relief mechanisms. The increased flexibility of the $\rho - relief\ LB$s is due to the dual-parameter control : the token pool capacity and the instantaneous relief.

The study has shown that the token pool capacity control does not seem to increase the efficiency achieved by setting the token pool capacity infinite and optimizing with respect to the instantaneous relief. As a result, it seems reasonable to focus on spacer-controller type traffic regulators, as it has been considered in some recent work[17,18,20], and seriously question the credit-based approach to congestion avoidance.

Finally, the $\sigma - relief\ TR$ has been introduced as a practical approach to implementing a $\rho - relief\ LB$ with infinite token pool. This mechanism corresponds to a near-periodic spacer which delivers a relief rate arbitrarily close to σ. The cell loss probabilities induced at the user premises by the $\sigma - relief\ TR$ were derived by analyzing a finite capacity queueing system with service opportunities determined by a periodic pattern made up of $C + 1$ subframes; service is provided at the first slot of each subframe. The cell loss probabilities were obtained through the solution of linear equations whose dimensionality is linearly dependent on the queue capacity and the number of subframes.

APPENDIX A

In this appendix, we present some comments on the derivation of equations (4)-(8). In the following, let:

M mark the beginning of subframe u.

k be the number of cells arriving over subframe u.

j be the state of the Markov Chain of the source at time instant $M + T_u$, where T_u is the length of subframe u.

- Clearly, starting from any state $(i, n, u), 0 \leq u \leq C - 1$, the system cannot move to a renewal point at $M + T_u$. This is due to our selection of the renewal point as $(0, 0, 0)$ and that it can only be reached at the beginning of subframe 0. Thus, the time needed to reach a renewal point will be equal to T_u plus the additional time needed to reach a renewal point from time slot $M + T_u$ (the end of the current subframe). This additional time is given by $X(j, q, u + 1)$, where j is the state of the source's Markov chain at $M + T_u$, $u + 1$ is the index of the following subframe and q is the state of the cell queue at $M + T_u$. The state of the queue is determined by the pattern of arrivals over the $\underline{uth}$ subframe as follows:

  If the queue is empty at the beginning of the u subframe, then it will remain empty after the first time slot of the subframe. This is due to the fact that an arriving cell–if any–will be serviced by the released token at the first time slot. Thus, the queue state at the end of the subframe will be determined by the number of cells (k') arriving over the remaining $T_u - 1$ slots of the subframe. If $k' \geq Q$, then the queue will become full and a number of cells equal to $(k' - Q)$ will be lost. However, if $k' < Q$, then the queue state will be k' at the end of the subframe.

- Starting from state $X(i,0,C)$, at the beginning of the last subframe (C), then as explained above, the queue will be empty after the first time slot and the system moves to a renewal point at $M+T_C$ if $j=0$ and no arrivals occur over the remainder of the subframe ($k'=0$). In this case $X(i,0,C)=T_C$. If $j\neq 0$, then $M+T_C$ cannot be a renewal point and if no arrivals occur, the cell queue will become empty and the system state at $M+T_C$ will be $(1,0,0)$. The additional time (beyond T_C) which will be required until the next renewal point is given by $X(1,0,0)$. Also, if $k'>0$, then $M+T_C$ cannot be renewal point and the time till the next renewal point will be equal to $T_C+X(j,q,0)$. Again, the queue state(q) will be determined by the number of cell arrivals k' over the remainder of the subframe. The remaining equations can be explained in a similar way.

APPENDIX B

In the following, the system of linear equations (9) is presented. First, let $f^m(i,j,k)$ be the probability of the event $(i \stackrel{m}{\rightarrow} j, a_m = k)$.$f^m(i,j,k)$ can be computed recursively as follows:

- For $m=1$, clearly:

$$f^1(i,j,k) \quad = \Pr(i \stackrel{1}{\rightarrow} j, a_1 = k) = p(i,j)g(i,k) \quad i\epsilon S,\ j\epsilon S,\ k\epsilon[0,1]$$

where:

$$g(i,k) \quad = \Pr\{k \text{ cells are generated from state } i\}$$

- For $m>1$:

$$f^m(i,j,0) \quad = \textstyle\sum_{l=0}^{1} f^{m-1}(i,l,0)f^1(l,j,0)$$

- For $m>1,\ 1\leq k\leq m$:

$$f^m(i,j,k) \quad = \textstyle\sum_{l=0}^{1}[f^{m-1}(i,l,k)f^1(l,j,0)+f^{m-1}(i,l,k-1)f^1(l,j,1)]$$

By applying the expectation operator to the equations (4)-(8), the following equations can be obtained.

For $i\epsilon S,\ 0\leq u\leq C-1$:

$$\overline{X}(i,0,u)=T_u \quad + \quad \textstyle\sum_{j=0}^{1}\sum_{k=0}^{Q-1}\sum_{S=0}^{1}\sum_{l=0}^{1} f^1(i,s,l)f^{T_u-1}(s,j,k)\overline{X}(j,k,u+1)$$

$$+ \quad \textstyle\sum_{j=0}^{1}\sum_{k=Q}^{T_u-1}\sum_{S=0}^{1}\sum_{l=0}^{1} f^1(i,s,l)f^{T_u-1}(s,j,k)\overline{X}(j,Q,u+1)$$

$$\overline{X}(i,n,u)=T_u \quad + \quad \textstyle\sum_{j=0}^{1}\sum_{k=0}^{Q-n} f^{T_u}(i,j,k)\overline{X}(j,n+k-1,u+1)$$

$$+ \quad \textstyle\sum_{j=0}^{1}\sum_{k=Q-n+1}^{T_u} f^{T_u}(i,j,k)\overline{X}(j,Q,u+1)$$

For $i\epsilon S$:

$$\overline{X}(i,0,C) = T_C + \sum_{s=0}^{1}\sum_{l=0}^{1} f^1(i,s,l) f^{T_C-1}(s,1,0)\overline{X}(1,0,0)$$
$$+ \sum_{j=0}^{1}\sum_{k=1}^{Q-1}\sum_{S=0}^{1}\sum_{l=0}^{1} f^1(i,s,l) f^{T_C-1}(s,j,k)\overline{X}(j,k,0)$$
$$+ \sum_{j=0}^{1}\sum_{k=Q}^{T_C-1}\sum_{S=0}^{1}\sum_{l=0}^{1} f^1(i,s,l) f^{T_C-1}(s,j,k)\overline{X}(j,Q,0)$$

For $i\epsilon S$:

$$\overline{X}(i,1,C) = T_C + f^{T_C}(i,1,0)\overline{X}(1,0,0) + \sum_{j=0}^{1}\sum_{k=1}^{Q-1} f^{T_C}(i,j,k)\overline{X}(j,k,0)$$
$$+ \sum_{j=0}^{1}\sum_{k=Q}^{T_C} f^{T_C}(i,j,k)\overline{X}(j,Q,0)$$

For $i\epsilon S,\ n \geq 2$:

$$\overline{X}(i,n,C) = T_C + \sum_{j=0}^{1}\sum_{k=0}^{Q-n} f^{T_C}(i,j,k)\overline{X}(j,n+k-1,0)$$
$$+ \sum_{j=0}^{1}\sum_{k=Q-n+1}^{T_C} f^{T_C}(i,j,k)\overline{X}(j,Q,0)$$

Thus, the coefficients of the unknowns and the constants of the system of linear equations (9) are given as follows:

$$b(i,n,u) = T_u \quad ,i\epsilon S,\ 0 \leq n \leq Q,\ 0 \leq u \leq C$$

For $i_1\ \epsilon S,\ i_2 \epsilon S,\ 0 \leq u \leq C-1,\ 0 \leq n_2 \leq Q-1$:

$$a(i_1,0,u,i_2,n_2,u+1) = \sum_{s=0}^{1}\sum_{l=0}^{1} f^1(i_1,s,l) f^{T_u-1}(s,i_2,n_2)$$

$$a(i_1,0,u,i_2,Q,u+1) = \sum_{k=Q}^{T_u-1}\sum_{s=0}^{1}\sum_{l=0}^{1} f^1(i_1,s,l) f^{T_u-1}(s,i_2,k)$$

For $i_1 \epsilon S,\ i_2 \epsilon S,\ 1 \leq n_1,\ 0 \leq u \leq C-1$:

$$a(i_1,n_1,u,i_2,n_2,u+1) = f^{T_u}(i_1,i_2,n_2-n_1+1), \quad 1 \leq n_2 \leq Q-1$$

$$a(i_1,n_1,u,i_2,Q,u+1) = \sum_{K=Q-n_1+1}^{T_u} f^{T_u}(i_1,i_2,k)$$

For $i_1 \epsilon S,\ i_2 \epsilon S,\ 1 \leq n_2 \leq Q-1$:

$$a(i_1,0,C,1,0,0) = \sum_{s=0}^{1}\sum_{l=0}^{1} f^1(i_1,s,l) f^{T_C-1}(s,1,0)$$

$$a(i_1,0,C,i_2,n_2,0) = \sum_{s=0}^{1}\sum_{l=0}^{1} f^1(i_1,s,l) f^{T_C-1}(s,i_2,n_2)$$

$$a(i_1,0,C,i_2,Q,0) = \sum_{k=Q}^{T_C-1}\sum_{s=0}^{1}\sum_{l=0}^{1} f^1(i_1,s,l) f^{T_C-1}(s,i_2,k)$$

For $i_1 \epsilon S$:

$$a(i_1,1,C,1,0,0) = f^{T_C}(i_1,1,0)$$

For $i_1 \epsilon S,\ i_2 \epsilon S,\ n_1 \geq 1$:

$$a(i_1, n_1, C, i_2, n_2, 0) = f^{T_C}(i_1, i_2, n_2 - n_1 + 1), \quad 1 \leq n_2 \leq Q - 1$$

$$a(i_1, n_1, C, i_2, Q, 0) = \Sigma_{K=Q-n_1+1}^{T_C} f^{T_C}(i_1, i_2, k)$$

APPENDIX C

In the following, we present the derivation of the cell loss equations:
For $i \epsilon S,\ 0 \leq u \leq C - 1$:

$$w(i, 0, u) = \left[\begin{array}{ll} w(j, k', u+1) & if\ (i \xrightarrow{1} m, a_1 = 0, 1), \\ & (m \xrightarrow{T_u - 1} j, a_{T_u - 1} = k'), 0 \leq k' \leq Q - 1 \\ k' - Q + w(j, Q, u+1) & if\ (i \xrightarrow{1} m, a_1 = 0, 1), \\ & (m \xrightarrow{T_u - 1} j, a_{T_u - 1} = k'), Q \leq k' \leq T_u - 1 \end{array} \right.$$

For $i \epsilon S,\ 0 \leq u \leq C - 1,\ n \geq 1$:

$$w(i, n, u) = \left[\begin{array}{ll} w(j, n+k-1, u+1) & if\ (i \xrightarrow{T_u} j, a_{T_u} = k), \\ & 0 \leq k \leq Q - n \\ k - Q + n - 1 + w(j, Q, u+1) & if\ (i \xrightarrow{T_u} j, a_{T_u} = k), \\ & Q - n + 1 \leq k \leq T_u \end{array} \right.$$

For $i \epsilon S$:

$$w(i, 0, C) = \left[\begin{array}{ll} 0 & if\ (i \xrightarrow{1} m, a_1 = 0, 1), \\ & (m \xrightarrow{T_C - 1} 0, a_{T_C - 1} = 0) \\ w(1, 0, 0) & if\ (i \xrightarrow{1} m, a_1 = 0, 1), \\ & (m \xrightarrow{T_C - 1} 1, a_{T_C - 1} = 0) \\ w(j, k', 0) & if\ (i \xrightarrow{1} m, a_1 = 0, 1), \\ & (m \xrightarrow{T_C - 1} j, a_{T_C - 1} = k'), 1 \leq k' \leq Q - 1 \\ k' - Q + w(j, Q, 0) & if\ (i \xrightarrow{1} m, a_1 = 0, 1), \\ & (m \xrightarrow{T_C - 1} j, a_{T_C - 1} = k'), Q \leq k' \leq T_C - 1 \end{array} \right.$$

For $i \epsilon S$:

$$w(i, 1, C) = \left[\begin{array}{ll} 0 & if\ (i \xrightarrow{T_C} 0, a_{T_C} = 0) \\ w(1, 0, 0) & if\ (i \xrightarrow{T_C} 1, a_{T_C} = 0) \\ w(j, k, 0) & if\ (i \xrightarrow{T_C} j, a_{T_C} = k),\ 1 \leq k \leq Q - 1 \\ k - Q + w(j, Q, 0) & if\ (i \xrightarrow{T_C} j, a_{T_C} = k),\ Q \leq k \leq T_C \end{array} \right.$$

For $i \epsilon S, n \geq 2$:

$$w(i,n,C) = \begin{cases} w(j,n+k-1,0) & if\ (i \stackrel{T_C}{\rightarrow} j, a_{T_C} = k), \\ & 0 \leq k \leq Q-n \\ k-Q+n-1+w(j,Q,0) & if\ (i \stackrel{T_C}{\rightarrow} j, a_{T_C} = k), \\ & Q-n+1 \leq k \leq T_C \end{cases}$$

The above equations may be explained in the following way: The total cell losses over $X(i,n,u)$ is equal to those lost over the first subframe of the session $X(i,n,u)$ plus those lost over the session $X(\cdot,\cdot,\cdot)$ initiated in the following subframe. The latter are given by $w(\cdot,\cdot,\cdot)$. The equations for the last subframe C can be explained similarly. By applying the expectation operator to the above equations, we obtain the following system of linear equations:

$$w(i_1,n_1,r_1) = b_w(i_1,n_1,r_1) + \sum_{i_2 \epsilon \Omega} \sum_{n_2 \epsilon S^Q} \sum_{r_2 \epsilon S^n} a(i_1,n_1,r_1,i_2,n_2,r_2).\overline{w}(i_2,n_2,r_2)$$

where:

For $i_1 \epsilon S, 0 \leq u \leq C$:

$$b_w(i_1,0,u) = \sum_{j=0}^{1} \sum_{s=0}^{1} \sum_{k=Q+1}^{T_u-1} \sum_{l=0}^{1} (k-Q) f^1(i_1,s,l) f^{T_u-1}(s,j,k)$$

For $i_1 \epsilon S, n_1 \geq 1, 0 \leq u \leq C$:

$$b_w(i_1,n_1,u) \quad = \quad \sum_{j=0}^{1} \sum_{k=Q-n_1+1}^{T_u} \quad (k-Q+n_1-1) f^{T_u}(i_1,j,k)$$

and $a(i_1,n_1,r_1,i_2,n_2,r_2)$ are as shown in Appendix B.

ACKNOWLEDGEMENTS

Research supported by the National Science Foundation under grant NCR-9011962.

REFERENCES

1. J. Bae, T. Suda, "Survey of Traffic Control Schemes and Protocols in ATM Networks," Proc. IEEE, (79), 2, Feb. 1991.
2. J. Turner, "New Directions in Communications (or Which Way to the Information Age?)," IEEE Commun. Mag., Oct. 1986.
3. M. Butto, E. Cavallero, A. Tonietti, "Effectiveness of the "Leaky Bucket" Policing Mechanism in ATM Networks," IEEE J. Select. Areas Commun., (9), 3, Apr. 1991.
4. M. Sidi, W. Liu, I. Cidon, I. Gopal, "Congestion Control Through Input Traffic Rate Regulation," IEEE Globecom'89 Conf.
5. E. Rathgeb, "Modeling and Performance Comparisons at Policing Mechanisms for ATM Networks," IEEE J. Select. Areas Commun., (9), 3, Apr. 1991.
6. "Congestion Control in High Speed Networks," Special Issue, IEEE Commun. Mag., Oct. 1991.

7. "B-ISDN: High Performance Transport," Special Issue, IEEE Commun. Mag., Sept. 1991.
8. W. Leland, "Window Based Congestion Management in Broadband ATM Networks : The Performance of Three Access-Control Policies," IEEE Globecom'89 Conf.
9. K. Sohraby, M. Sidi, "On the Performance of Bursty and Correlated Sources Subject to Leaky Bucket Rate-based Access Control Scheme," IEEE Infocom'91 Conf.
10. D. Hughes, H. Bradlow, "Congestion Control in ATM Network", ITC-13, pp. 835-840, 1991, North-Holland.
11. F. Guillemin, A. Dupuis, "A basic Requirment for the Policing Function in ATM Networks", Computer Networks and ISDN Systems, pp. 311-320, Vol. 24, 1992.
12. H. Ahmadi, R. Guerin, K.Sohraby, "Analysis of a Rate-Based Access Control Mechanism for High-Speed Networks," IEEE Globecom'90 Conf.
13. Guoliang Wu and Jon W Mark, "Discrete Time Analysis Of Leaky Bucket Congestion Control", ICC'92 Conf., pp. 1196-1200, June 14-18, Chicago.
14. K.Q.Liao, Z. Dziong, L.Mason, N. Tetreault, "Effectiveness of The Leaky Bucket Policing Mechanism", ICC'92 Conf., pp. 1201-1205, June 14-18, Chicago.
15. I. Stavrakakis, "A Two-Server Queueing System with Periodic and Credit-Based Server Availabilities", submitted to the Annals of Operations Research.
16. G. Gallasi, G. Rigolio, L. Fratta, "ATM: Bandwidth Assignment and Bandwidth Enforcement Policies", IEEE Infocom'91, pp. 1788-1793, 1991.
17. K. Bala, I. Cidon, K. Sohraby, "Congestion Control for High Speed Packet Switched Networks," IEEE ICC'90 Conf.
18. P. Boyer, "A Congestion Control for ATM," Int. Teletraf. Cong. Sem, Oct. 1990, New Jersey.
19. B. Lague, C. Rosenberg, F. Guillemin, "A Generalization of Some Policing Mechanisms", IEEE Infocom'92, pp. 767-775, May 6-8, Florence, Italy.
20. F. Guillemin, P. Boyer, A. Dupuis, L. Romoeuf, "Peak Rate Enforcement in ATM Networks", IEEE Infocom'92, pp. 753-758, May 6-8, Florence, Italy.
21. G. Rigolio, L. Fratta, "Input Rate Regulation and Bandwidth Assignment in ATM Networks: an Integrated Approach", ITC-13, pp. 141-146, 1991, North-Holland.

SPACE PRIORITY BUFFER MANAGEMENT FOR ATM NETWORKS

David Tipper, Srinivas Pappu, Aaron Collins, and John George

Electrical and Computer Engineering Department
Clemson University
Clemson, SC 29634-0915

INTRODUCTION

Broadband Integrated Services Digital Networks (BISDN) will support a wide variety of services such as voice, data, and video applications utilizing a limited set of connection types and user network interfaces. Standards committees have decided on Asynchronous Transfer Mode (ATM) as the target transport technique for BISDN. ATM is a connection oriented transport technique in which all information is conveyed using a fixed size packet (called a "cell") for switching and transmission purposes so as to dynamically share the network resources between connections. Since BISDN's will support various classes of multimedia traffic with different bit rates and quality of service requirements, traffic control and resource management are crucial in order to guarantee a grade of service. Several mechanisms will exist in a BISDN to control traffic, such as call admission control, input rate regulation and routing. Here we focus on a particular type of congestion control dealing with packet discarding at ATM buffers in order to guarantee a specified cell loss rate.

It is well known that for a limited buffer system supporting different classes of traffic, such as an ATM queue at a switch or multiplexor, efficient buffer management schemes are necessary to minimize loss rates. One mechanism for buffer management is the introduction of space priorities among the incoming traffic. Space priority implies that higher priority cells are favored in receiving space in the buffer at a queue. Standards bodies have agreed upon reserving one bit in the ATM cell header to indicate space priority [1]. When congestion occurs at a BISDN queue lower priority cells can be discarded in order to insure a smaller cell loss rate for higher priority cells. Strategies for determining which cells to discard when congestion occurs are termed space priority buffer management schemes or priority cell discarding schemes in the literature. Under the proposed ATM standards space priority may be implemented in the network at one of two levels, either the connection level or the cell level.

In the case of connection level implementation all cells within a specific connection (i.e. virtual circuit) are given the same space priority which is determined by the type of traffic in the connection. For example, telephone can tolerate some loss (e.g. 10^{-6}) with acceptable reproduction still possible and could be marked low priority, whereas some types of data traffic may require very low loss rates (e.g.

Asynchronous Transfer Mode Networks, Edited by Y. Viniotis
and R.O. Onvural, Plenum Press, New York, 1993

10^{-8}) and should be given high priority. The space priority would be determined by the connection acceptance algorithm at the time of call set-up, depending on the grade of service required by the connection.

In contrast, for cell level implementation the the priority can be marked differently for each cell in an individual connection. The priority can be determined in a number of ways with the priority of the cells being marked either by the traffic source itself or by a bandwidth enforcement device. One approach is to have the traffic source, such as a variable rate subband video encoder, mark the cells on the basis of the information contained in the cell as either essential cells (high priority) or nonessential cells (low priority). An alternate method to cell marking is to have a bandwidth enforcement device such as a virtual leaky bucket determine the priority on the basis of specified traffic parameters. For example, the user negotiates a specified mean rate, peak rate and burst length with the network and as long as the user stays within the negotiated parameters the virtual leaky bucket tags cells as high priority. When the user exceeds the negotiated parameters the cells violating the parameters are marked as low priority and suffer a higher cell loss rate.

Regardless of the implementation of the space priority, the problem at a ATM queue buffer is the same, namely: how to manage the buffer space to satisfy the loss requirements while maximizing the potential throughput. Here we propose two new space priority schemes based on cell pushout, wherein, a cell can overwrite or pushout a cell already in the buffer. The performance of the new pushout schemes is compared to existing space priority schemes using a single server finite capacity queueing model representation of an ATM multiplexer or output buffer of an ATM switch. In the queueing model the traffic is grouped into two classes: non-real time traffic, such as a file transfer, which is sensitive to the cell loss rate and real time traffic, such as voice, which is delay sensitive but can tolerate a higher loss rate. The performance characteristics of the schemes are studied for a wide range of traffic parameters and the superiority of the two pushout schemes demonstrated. The results are determined using an analytical model for the case of the cells arriving according to Bernoulli processes.

SPACE PRIORITY MECHANISMS

A considerable amount of work exists on the topic of buffer management for packet switched networks[2-4] and more recently on space priority schemes in particular[5-12]. Early work[2,3] focused on the problem of how to share a finite buffer space among a set of separate queues at a network node. A notable study of this type is that by Thareja and Agrawala [4], where they noted that the common practice of placing a limit on the amount of buffers that can be used by any queue will result in a wastage of buffer space as traffic may be discarded even though the total buffer space is not full. They suggested delaying the decision to discard packets until the buffer is totally full since there is a chance that the packets maybe admitted and serviced prior to the buffer filling. A packet upon arriving to a full buffer could possibly overwrite or pushout a packet from another queue depending on some queue length thresholds. This type of buffer management scheme was termed a delayed resolution scheme or a pushout scheme and was analyzed for the case of Possion arrivals and exponential service.

Recently buffer management work has focused on analyzing the management of multiple classes of traffic at a single queue and various space priority mechanisms have been proposed. Some of the schemes proposed include separate buffers for each class[10,11], partial buffer sharing (also called nested threshold) [5-10] and pushout schemes[10–13]. In the seperate buffer scheme each class of traffic has a dedicated buffer and cells are discarded only when the buffer for that class is full. Hence, no special buffer management is required. However, the space priority processing must be implemented at the connection level. Also, when compared to the other schemes, a larger total buffer space is needed to meet a specified set of cell loss requirements since no buffer sharing takes place.

In the partial buffer sharing scheme a common buffer is provided for all classes and the sharing of the buffer is controlled by a set of discarding thresholds. Let T_i denote the discard threshold for the class i traffic and assume that class i has priority over class $i+1$. Buffer access for class i cells is granted if less than T_i buffer spaces are occupied. The class with the highest priority is given access to the entire buffer (i.e. $T_i = K$ where K is the total buffer space). Various cell loss requirements for different loading scenarios can be provided by adjustment of the threshold values. Note that there is some buffer wastage in this scheme since lower priority cells can be discarded even though the entire buffer space is not occupied. This has motivated the development of pushout schemes where buffer access is always granted as long as the entire buffer space in not full. When the buffer is full an arriving cell can overwrite any lower priority cells in the buffer, if no lower priority cells are present, the cell is discarded.

The performance of the three types of space priorityschemes has been analyzed in a series of studies for various ATM traffic source models (e.g. Bernoulli, Poisson, etc...) and no one scheme has proven to be uniformly best. However, we feel that a drawback of the current literature is that the proposed pushout schemes do not allow any adjustment (i.e. tuning) of the cell loss rates among the various classes. Specifically, in the previous studies with two classes of traffic (the primary case of interest) class 1 cells (high priority) always overwrite class 2 cells (low priority). Hence, class 1 traffic may be provided a grade of service exceeding what is required for the given traffic pattern and cell loss requirements. Here we propose two new pushout schemes which provide for adjustment of the cell loss rates among the traffic classes.

The two schemes are termed the threshold pushout and P_{ow} pushout strategies and are defined in the following for the case of two classes of traffic. For both strategies all cells are admitted until the common buffer space is full. In the threshold pushout scheme the space priority is determined according to a set of overwrite thresholds $[T_1, T_2]$ where we require that the sum of the thresholds equal the total buffer space (i.e. $T_1 + T_2 = K$). A class 1 cell arriving to a full buffer can overwrite a class 2 cell if the number of class 2 cells in the buffer exceeds the T_2 threshold, otherwise the class 1 cell is discarded. Similarly, a class 2 cell arriving to a full buffer can pushout a class 1 cell if the number of class 1 cells exceeds the T_1 threshold, otherwise the class 2 cell is discarded. Note that $T_1 = K - T_2$ and by adjustment of T_2 one can vary the cell loss rates provided the two classes. The P_{ow} pushout scheme is similar except the space priority is determined by an overwrite probability P_{ow}. Specifically, a class 1 cell arriving to a full buffer will overwrite a class 2 cell with probability P_{ow}. Conversely, a class 2 cell is granted access to a full buffer via pushing out a class 1 cell with probability $1 - P_{ow}$. Clearly, varying P_{ow} provides a mechanism for adjusting the cell loss rates of the two classes.

PERFORMANCE ANALYSIS

Utilizing a queueing analysis, the performance of the threshold pushout and P_{ow} pushout strategies was compared to the partial buffer sharing and no priority

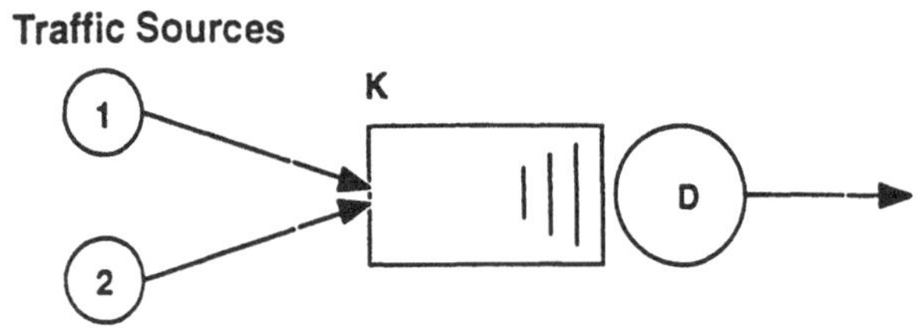

Traffic Sources

K

1

2

D

Figure 1. ATM buffer queueing model

(i.e. buffer space allocated on a first come basis) schemes. Consider the queueing model of an ATM buffer illustrated in Figure 1, where two classes of traffic join a finite buffer of size K and require a constant service time of D. In the analysis that follows was assume that time is divided into slots of constant length equivalent to one cell transmission time. Following Petr[7], the arrival process for each class of traffic is modeled as Bernoulli with p_i denoting the probability of a class i cell arrival in a slot.

The state of the queueing system is denoted by the two tuple (i, j) where i represents the number of class 1 cells in the system and j is the number of class 2 cells in the system. Note, that observing the state of the system at the slot times defines a discrete time Markov chain. The state probabilities are defined as $\pi_{i,j}$ = prob {i class '1' cells, j class '2' cells}. Similarly, the vector π_n of state probabilities repsenting those states where n cells are in the system is defined as $\pi_n = [\pi_{0,n}, \pi_{1,n-1}, \ldots, \pi_{n,0}]$. Lastly, we denote the entire set of state probabilities by the vector π which is given by $\pi = [\pi_0, \pi_1, \pi_2, \ldots, \pi_K]$.

The steady state behavior of the queueing system is described by the state probabilities π, from which one can determine performance measures such as the cell loss rates, and average queue lengths. It is well known that for a discrete time Markov chain the steady state probabilities π are determined by the solution of $\pi = \pi \cdot \mathbf{P}$ and the normalization condition $\pi \cdot \mathbf{e} = 1$, where $\mathbf{P}$ is the state transition matrix and $e = [1, 1, 1, \ldots, 1]^T$. Note, that if the number of cells in the system is greater than zero then a cell will depart at the end of each slot time and at most two cells arrivals can occur in any given slot, which gives rise to a state transition matrix $\mathbf{P}$ of the form shown below.

$$
\mathbf{P} = \begin{bmatrix}
\alpha_{00} & \alpha_{01} & \alpha_{02} & 0 & 0\ldots & 0 & 0 \\
\alpha_{10} & \alpha_{11} & \alpha_{12} & 0 & 0\ldots & 0 & 0 \\
0 & \alpha_{21} & \alpha_{22} & \alpha_{23} & 0\ldots & 0 & 0 \\
\vdots & \vdots & \vdots & \vdots & \vdots & \ddots & \vdots \\
0 & 0 & 0 & 0 & 0\ldots & \alpha_{K,K-1} & \alpha_{K,K}
\end{bmatrix}
$$

Where the α_{ij} in the state transition matrix are submatrices defining the rate of transition from state i to state j. The exact form of the submatrices α_{ij} depends on the state (i, j) and space priority scheme modeled. The precise form of α_{ij} in terms of the state and arrival probabilities p_i for each scheme is given in Pappu[14]. An analytical expression for the state probabilities π is difficult to obtain for the pushout schemes. However, one can easily solve for the state probabilities using standard numerical matrix solution techniques.

In order to validate the analytical queueing model, a corresponding simulation model was developed and the results compared. The simulation model was developed in C and the standard method of replications with deletion of the initial transient was used to statistically analyze the output data. The simulation was run until 90% confidence intervals with 1% relative precision were obtained on all performance measures. An extensive set of experiments comparing the throughput, delay, cell loss rates, and queue lenghts of the two models is given in Pappu[14] with representative results reported here.

The system of Figure 1 was studied for a fixed buffer space of seven ($K = 7$) with the arrival rate of class 1 cells set at $p_1 = .5$ and the arrival rate of class 2 cells varied over the range $0 \leq p_2 \leq 1$ in order to construct performance curves. Typical results of the experiment are shown in Figures 2, 3, and 4. In the figures the simulation values are denoted by symbols (e.g. Δ) whereas the continuous curves represent the analytical model results. One can see that there is good agreement between the analytical and simulation results.

Figure 2 shows the behavior of the number of cells in the queueing system for the partial buffer sharing scheme as the load and discarding thresholds are varied. In Figure 2, the discarding threshold values corresponding to a particular curve are denoted in brackets (i.e. $[T_1, T_2]$). Note that the curve for the partial buffer sharing scheme with threshold levels [7,7] corresponds to the results for the no priority scheme and both pushout schemes (regardless of their threshold or overwrite probability settings).

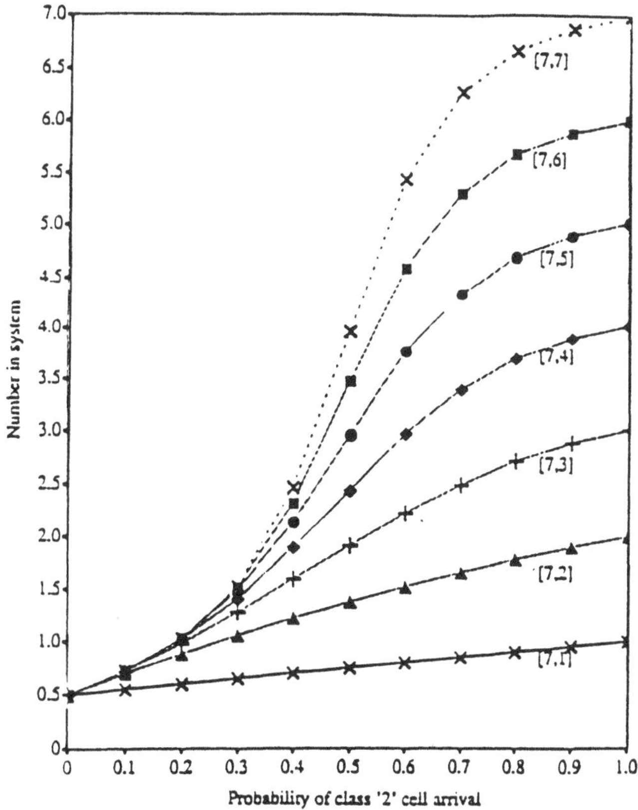

Figure 2. Behavior of number in the system for partial buffer sharing.

In Figure 3 the breakdown of the number of cells in the system into class 1 and class 2 for the partial buffer sharing scheme is illustrated. From Figures 2 and 3 one can clearly see that as the threshold for class 2 is increased thereby allowing more buffer space to be shared, the number of class 2 and class 1 cells queued increases.

Figure 4 shows a similar breakdown of the number of class 1 and class 2 cells in the system for the pushout threshold scheme as the load and pushout thresholds are varied. One can clearly see, that the total number in the system is independent of the threshold values, but that the percentage of mixture class 1 and class 2 cells is determined by the overwrite thresholds. It should be noted, that the [0,7] and [7,0] threshold settings correspond to the P_{ow} pushout scheme with settings $P_{ow} = 0$ and $P_{ow} = 1$ respectively. Also, Pappu[14] has shown that by varying P_{ow} one can get a set of curves for any point in the space bounded by the [0,7] and [7,0] curves of Figure 4.

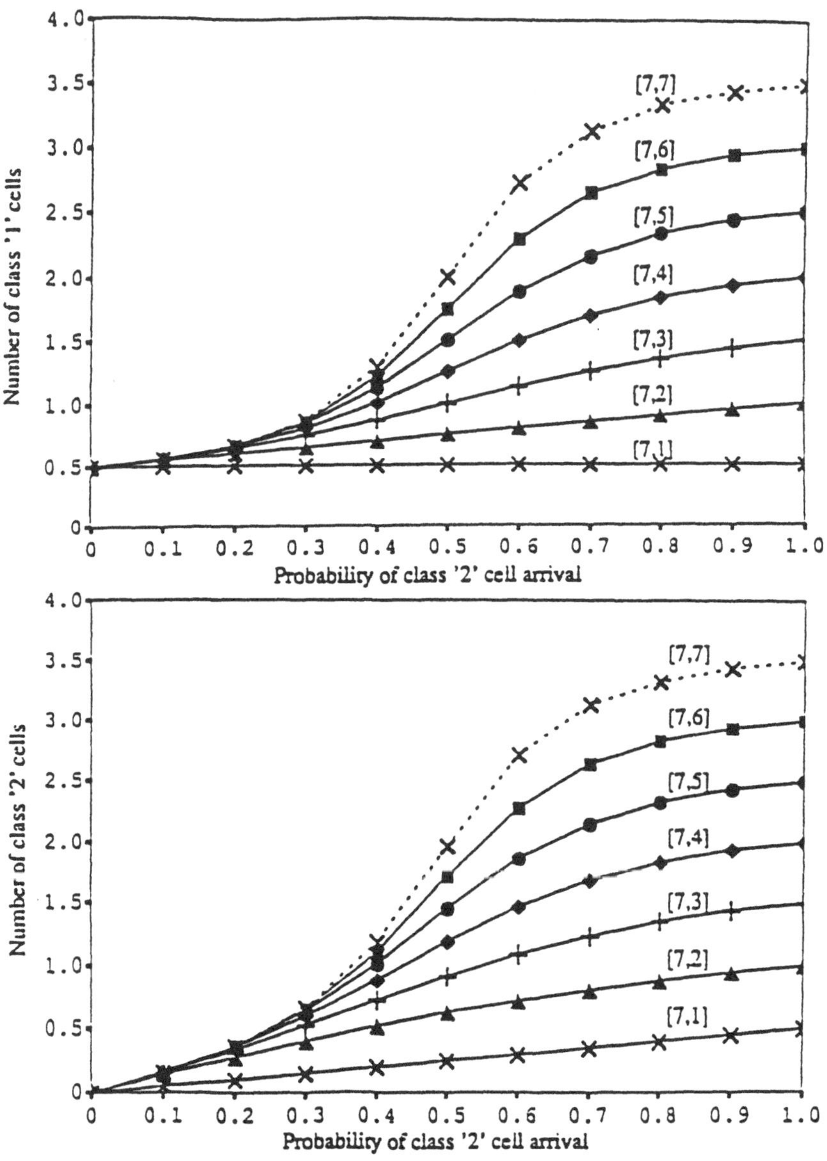

Figure 3. Behavior of the number in the system for each class for partial buffer sharing.

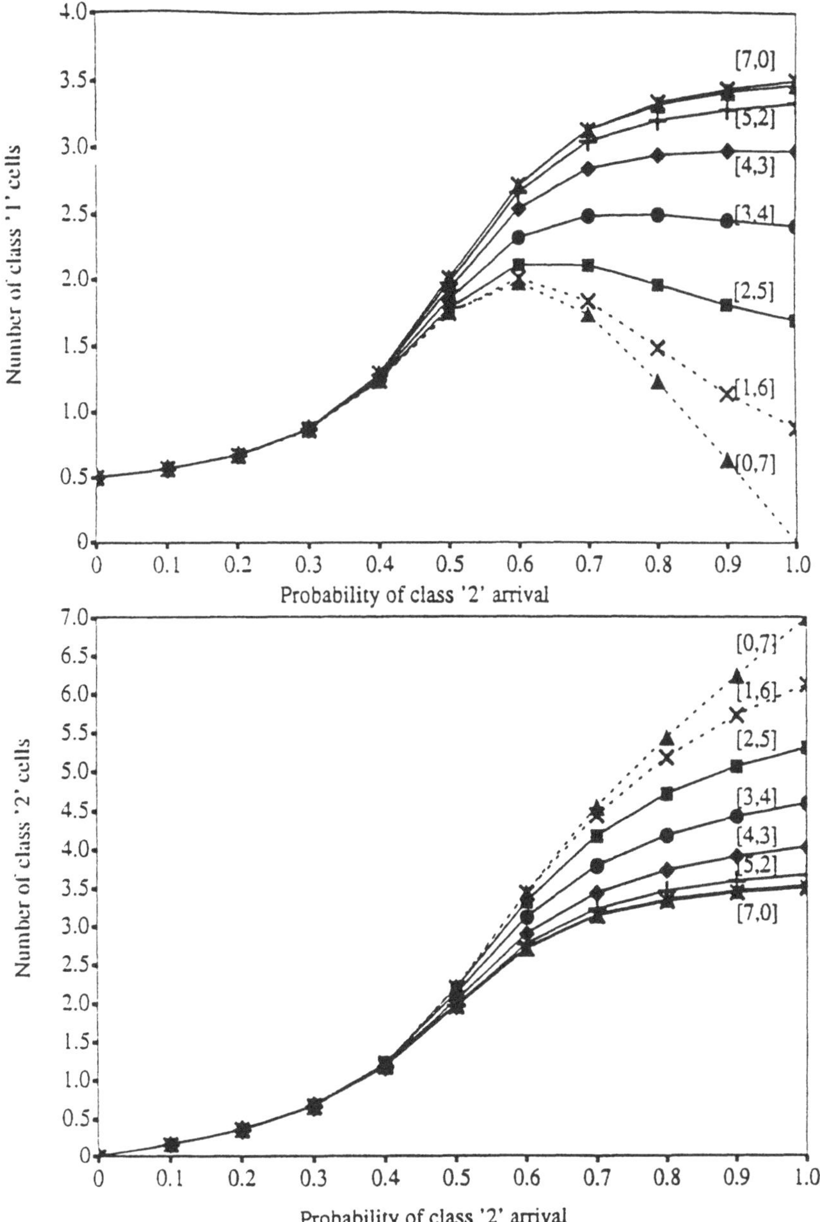

Figure 4. Behavior of the number in the system for each class for the threshold pushout scheme.

Having validated the analytical models a study was conducted to determine which scheme provided the best performance. Let Δ_i denote the maximum acceptable cell loss rate for class i traffic. Given a set of cell loss requirements (Δ_1, Δ_2) and the traffic mixture of the load (i.e. (% class 1, % class 2)), the total load was varied until the maximum offered load (MOL) that could be supported while still satisfying the loss requirements was determined. Tables 1 and 2 show typical results of this study for the two sets of cell loss requirements most often cited in the literature. Note that several traffic mixtures were considered for each set of loss

Table 1. Maximum offered load for various traffic mixtures with cell loss requirements $(10^{-10}, 10^{-6})$.

$\Delta_1 = 10^{-10}$	MOL for (% class 1 , % class 2) traffic mix								
$\Delta_2 = 10^{-6}$	10,90	20,80	30,70	40,60	50,50	60,40	70,30	80,20	90,10
No Priority	0.43052	0.35350	0.32219	0.30939	0.30829	0.31776	0.34042	0.38569	0.48840
Partial Buffer [7,6]	0.67295	0.55555	0.49824	0.46794	0.45435	0.45439	0.46928	0.50654	0.59414
Pushout Threshold	0.73945	0.62802	0.57086	0.54019	0.52667	0.52749	0.54407	0.58386	0.67303
Pushout Pow	0.73945	0.62802	0.57086	0.54019	0.52667	0.52749	0.54407	0.58386	0.67303
% improvement over Partial Buffer	9.88%	13.04%	14.57%	15.44%	15.92%	16.08%	15.94%	15.44%	13.67%

Table 2. Maximum offered load for various traffic with cell loss requirements $(10^{-8}, 10^{-6})$.

$\Delta_1 = 10^{-8}$	MOL for (% class 1 , % class 2) traffic mix								
$\Delta_2 = 10^{-6}$	10,90	20,80	30,70	40,60	50,50	60,40	70,30	80,20	90,10
No Priority	0.55629	0.46859	0.43140	0.41604	0.41491	0.42668	0.45424	0.50772	0.62124
Partial Buffer [7,6]	0.67295	0.55555	0.49824	0.46794	0.45435	0.45439	0.46928	0.50654	0.59414
Pushout Threshold	0.73947	0.62802	0.57088	0.54023	0.52675	0.52766	0.54441	0.58465	0.67541
Pushout Pow	0.73945	0.62802	0.57086	0.54019	0.52667	0.52749	0.54407	0.58386	0.67303
% improvement over Partial Buffer	9.88%	13.04%	14.58%	15.44%	15.93%	16.12%	16.00%	15.45%	13.67%

requirements. In the results listed for the partial buffer sharing and pushout schemes only the best case is given (i.e. the threshold values and overwrite probabilities were varied until the largest MOL was found). From the tables one can clearly see that the two pushout schemes proposed here provide a significant improvement in potential throughput over the partial buffer sharing scheme. This clearly contradicts the prevailing literature where only a slight improvement is to be expected from a pushout scheme. This is primarly due to the fact that the optimum pushout scheme for the cases studied was usually not the case of having class 1 traffic always overwrite class 2 traffic. Also, note that a software emulation of a board level hardware implementation of the pushout threshold scheme for a buffer of size 256 cells has been developed. The implementation is done entirely in hardware and the FIFO ordering of packets at the output of the buffer is preserved by using link lists as pointers for reading out packets. The hardware design was shown to easily run at speeds up to 75 Mbps and when compared to a hardware implementation of a partial buffering scheme provided about a 15% improvement in throughput.

SUMMARY

In this paper we have proposed two new pushout space priority schemes for ATM networks namely: threshold pushout and P_{ow} pushout. The distinguishing advantage of these new schemes is the ability to adjust the cell loss rates

between the traffic classes. The superior performance of the pushout schemes when compared with partial buffer sharing was demonstrated. Specifically it was shown that for a specified buffer size, traffic mixture and set of cell loss requirements the pushout schemes could support a significantly higher maximum load.

REFERENCES

1. CCITT Draft Recommendation I.361, "ATM layer specification for B-ISDN," *Study Group XVII*, Geneva, Switzerland, (1990).
2. M. Ireland, "Buffer management in a packet switch," *IEEE Transactions on Communications*, Vol. Com-26, pp. 328-337, (1978).
3. F. Kamoun and L. Kleinrock,"Analysis of shared finite storage in computer network node environment under general traffic conditions," *IEEE Transactions on Communications*, Vol. Com-28, pp. 992-1003, (1980).
4. A. Thareja and A. Agrawala,"On the design of optimal policy for sharing finite buffers," *IEEE Transactions on Communications*, Vol. Com-32, pp. 737-740, (1984).
5. D. Petr, L. DaSilva, and V. Frost, "Priority discarding of speech in integrated packet networks," *IEEE Journal on Selected Areas on Selected Areas in Communications*, Vol. 7, pp. 644-656, (1989).
6. D. Petr and V. Frost, "Priority cell discarding for overload control in B-ISDN/ATM networks: An analysis framework," *International Journal on Digital and Analog Communication Systems*, (1990).
7. D. Petr and V. Frost, "Nested threshold cell discard for ATM overload control: optimization under cell loss constraints," *Proceedings of IEEE INFOCOM '91*, Bal Habour, FL, (1991).
8. V. Yau and K. Pawlikowski, "Improved nested-threshold cell discard buffer management mechanisms," *Proceedings of IEEE Tencon '92*, Melbourne, Australia, (1992).
9. S. Sumita,"Synthesis of an output buffer management scheme in a switching system for multimedia communications," *Proceeding of IEEE INFOCOM '90*, San Francisco, CA. pp. 1226-1233, (1990).
10. G. Hebuterne and A. Gravey, "A space priority queuing mechanism for multiplexing ATM channels," *Computer Networks and ISDN Systems*, Vol. 20, pp. 37-43,(1990).
11. H. Kroner, "Comparative performance study of space priority mechanisms for ATM networks," *Proceeding of IEEE INFOCOM '90*, San Francisco, CA. pp. 1136-1143, (1990).

12. H. Kroner, G. Hebuterne, P. Boyer and A. Gravey,"Priority management in ATM switching nNodes," *IEEE Journal on Selected Areas in Communications*, Vol. 9, pp. 418-427, (1991).
13. A. Gravey and G. Hebuterne, "Mixing time and loss priorities in a single queue," *Proceedings of 13th International Teletraffic Congress*, Copenhagen, Denmark, pp. 147-152, (1991).
14. S. Pappu, "Analysis of buffer management space priority mechanisms in broadband ISDNS," M.S. thesis, Clemson University, Clemson, SC, (1992)

ASSIGNABLE GRADE OF SERVICE USING TIME DEPENDENT PRIORITIES -- N CLASSES

Mark Perry,[1] Chandramouli Sargor,[1] and Arne Nilsson[2]

[1]BNR, 35 Davis Drive
Research Triangle Park, North Carolina
[2]CCSP, Department ECE
North Carolina State University
Raleigh, North Carolina

ABSTRACT

A new generation of High Speed Switching equipment that can handle multiple types of information -- such as voice, video and data is being developed. Because voice services dominate current telephone network applications, bandwidth for network access is based on Grade-of-Service (GOS) requirements for it. This paper provides a method for allocating access capacity for High Speed networks that carry multiple types of information with different GOS requirements. The GOS is assumed to be given in terms of mean waiting time until access to the network is provided. We will develop an algorithm that calculates priority factors which will meet given Grade-of-Service requirements for different types of information.

I. INTRODUCTION

The current telephone network carries predominately one type of information -- voice. Consequently the Grade-of-Service (GOS) for access to the network was chosen to fit voice applications. Often this was specified in terms of dial-tone delay, which is the time a caller waits for dial-tone. The dial-tone delay can be more generically described as mean-time to access the network.

With recent advances in high speed networking it is now possible to carry multiple types of information such as voice, video, data, etc. over the same facilities. With multiple types of traffic, the GOS for access to network should no longer be built around just voice. Rather, it should account for all types of traffic. It could be that we want to give voice traffic quicker access than video, and video quicker access than data. A natural way to distinguish between our traffic is to have high-priority information given network access before low-priority information. But the priorities should not be the only criteria for determining the service a traffic type receives. Regardless of the classification of information, it should not be denied service for too long. A kind of priority that accounts for both time spent waiting for service, and the priority of the information, is an Aging Factor. An Aging Factor is a real number, which along with the queuing time of a request for network access determines the priority of the request. For example, if we have both a data and a voice call queuing for access to the

Asynchronous Transfer Mode Networks, Edited by Y. Viniotis
and R.O. Onvural, Plenum Press, New York, 1993

high speed network, and they have queued for 10 and 5 seconds, and have aging factors of 3 and 7 respectively, then their priorities are (10x3=30) and (7x5=35).

In this paper -- for a given set of GOS requirements -- we will develop an algorithm that calculates the aging factors for various types of information and the necessary access capacity. Before a formal analytic description of the problem is given, we will first discuss access to high speed networks.

II. ACCESS TO A HIGH SPEED NETWORK

The purpose of this section is to describe prioritized access to high speed networks. For concreteness we will assume that delivery of information to the high speed network is packaged in the Synchronous Optical NETwork (SONET) [1] format. SONET was initiated by Bellcore and the Regional Bell Operating Companies in the Unites States, to provide a standard for interconnecting fiber systems at the optical level. The SONET format has a hierarchy of signals -- shown in Table 1.

TABLE 1 SONET SIGNAL HIERARCHY -- NORTH AMERICA

Electrical Signal	Optical Signal	Data Rate (Mbps)	CCITT Designation
STS-1	OC-1	51.84	
STS-3	OC-3	155.52	STM-1
STS-9	OC-9	466.56	STM-3
STS-12	OC-12	622.08	STM-4
STS-18	OC-18	933.12	STM-6
STS-24	OC-24	1244.16	STM-8
STS-36	OC-36	1866.24	STM-12
STS-48	OC-48	2488.32	STM-16

Let's examine one of the frames, say the STS-1.

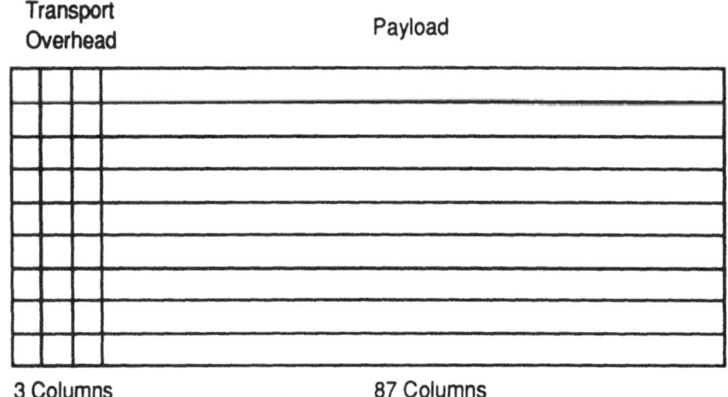

Figure 1: STS-1 frame format.

The frame has been broken up into rows (9) and columns (90), for illustrative purposes. Each cell contains a single byte. Since there are 8000 frames/second, the data rate is 51.84 Mbps.

There are many end-users that do not require the payload of a full STS-1 frame. Consequently the SONET standard supports partitioning the payload into Virtual Tributaries (VTs) -- frames at lower bit rates than STS-1. For example a DS1 (1.544 Megabits/sec) would be mapped into 3 of the payload columns of the STS-1 frame. Consequently the payload could hold 28 different DS1 signals. These VTs may be viewed as a set of finite resources that end-users compete for to gain access to the network. That is, if we assume we have 500 users that occasionally require a DS1 signal, and all of the DS1 signals are carried via a SONET signal, say a single STS-3, then we have competition for the 84 DS1 signals our STS-3 frame can carry. With this concentration it will be necessary to provide our end-users with a GOS for access to the VTs in the SONET payload. Since not all end-users applications are the same, some will require quicker access than others, it makes sense to have different access time GOSs for different end-users. Perhaps one end-user transmits a large amount of data every night that is not real time sensitive, he might want a GOS of a mean time to access the network of 30 minutes. But another end-user has real time sensitive data and needs quicker access, say mean time until access of 1 minute. These different GOSs for different end-users who are competing for the same 84 servers (VTs) can be accomplished by using aging factors. The following figure summarizes this discussion.

We now turn our attention to the formal problem description and its solution.

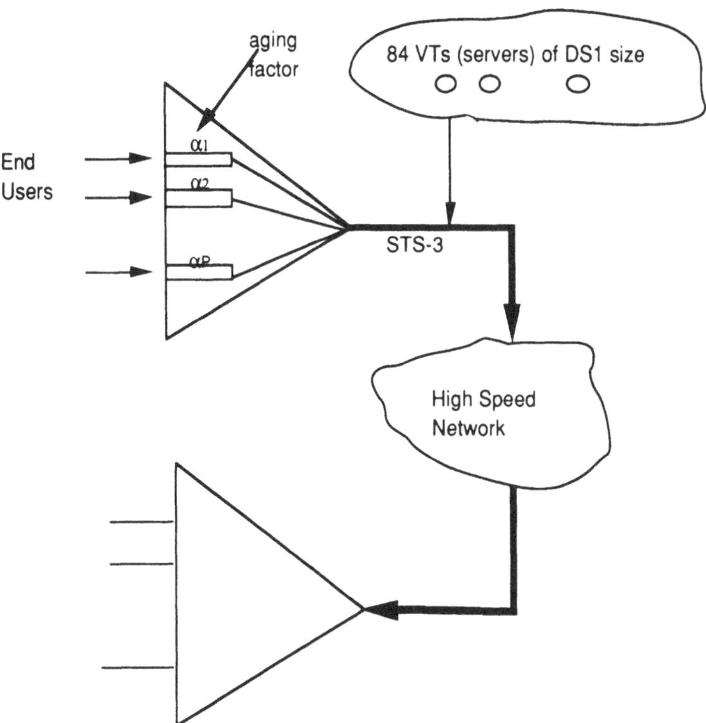

Figure 2: Access to a High Speed Network with VTs as Servers and Aging factors for priority.

III. PROBLEM DEFINITION

The system that we are interested in analyzing is depicted in Figure 3.

The following definitions will facilitate our discussion:

P	total number of classes
c	number of servers (Virtual Tributaries)
α_i	aging factor of class i $\quad 0<\alpha_1\leq\alpha_2\leq\ldots\leq\alpha_P$
λ_i	arrival rate of class i (Poisson)
x_i	mean service time of class i calls
ρ_i	traffic due to class i ($\lambda_i x_i$)
W_0	the initial expected waiting time an arbitrary customer experiences, if any, until the first epoch at which a server becomes available for service.
W_i	mean waiting time of class i
w_i	the waiting time ratio with respect to the lowest priority class W_i/W_1

Aging factors were used by Kleinrock in 1964 to determine the expected waiting times for single server systems[2][3]. Federgruen and Groenevelt investigated aging factors for multiple server systems [4]. This paper will extend their analysis and apply the result to the practical problem of allocating access capacity.

In order to allocate access capacity it is necessary to meet Grade of Service (GOS) requirements. The GOS is typically specified in terms of the waiting time requirements for the different classes. The number of VTs and the aging factors that will achieve these waiting times need to be determined. However, it is usually impossible to satisfy the absolute values of the waiting times in an implementation using aging factors alone. On the other hand, it is usually possible to satisfy the waiting time ratios for the specified GOS. For example, for the simple two class case we may want to achieve waiting times of 4 and 2. Since the number of servers, c, is an integer, c = 3 may lead to waiting times of 5 and 2.5 while c = 4 may result in waiting times of 3.8 and 1.9 for appropriate α's. Both of these waiting times satisfy the required 2:1 ratio, however neither of them meets the specified absolutes of 4 and 2. We therefore focus on the waiting time ratios.

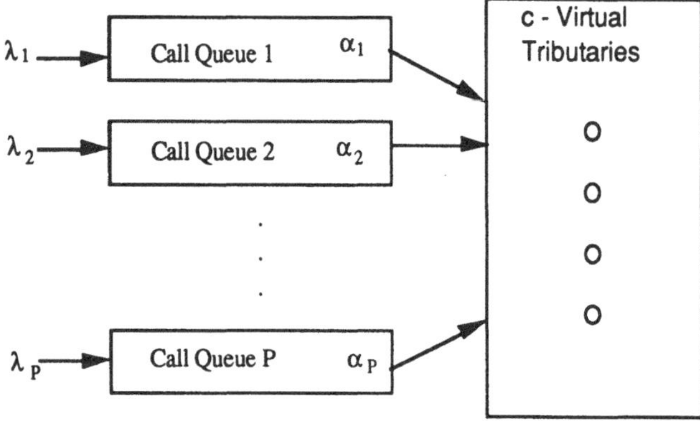

Figure 3: The queuing system with multi-class services.

IV. ANALYSIS

The problem that we are concerned with is the computation of the correct set of aging factors, given a *set of mean waiting times* (the GOS requirements), that will yield the desired *ratio of mean waiting times*. The waiting time ratios are defined with respect to the lowest class.

Given a set of aging factors $\{\alpha_i\}$, the computation of the waiting times $\{W_i\}$ is straightforward from the set of equations

$$W_p = W_0 + \frac{1}{c}\sum_{j=1}^{p-1} \rho_j W_j \frac{\alpha_j}{\alpha_p} + \frac{1}{c}\sum_{j=p}^{P} \rho_j W_j + \frac{1}{c}\sum_{j=p}^{P} \rho_j W_p (1 - \frac{\alpha_p}{\alpha_j}) \quad (1)$$

as outlined in [4].

Note, for multiple servers ($c>1$), equation (1) holds when the service times are exponentially distributed with common means. However, simulation results have shown this to be a good approximation when the common mean assumption is relaxed. Section V. gives some numerical results.

The reverse problem of obtaining the set of aging factors that will yield a desired set of waiting times is not trivial, because the equations that must be solved for the aging factors α_i are nonlinear. The equations above can be written in matrix form as

$$\mathbf{A}\,\mathbf{W} = \mathbf{e}\,(-W_0\, c) \quad (2)$$

where W is a column vector with elements W_i and **e** is a column vector of 1's. The elements of the matrix A are given by

$$\begin{aligned} A_{ii} &= \rho_i - D_i & \\ A_{ij} &= \rho_j (\alpha_j / \alpha_i) & i > j \\ A_{ij} &= \rho_j & i < j \end{aligned}$$

where D_i is defined as

$$D_i = c - \sum_{j=i+1}^{P} \rho_j (1 - \frac{\alpha_i}{\alpha_j})$$

If we are interested only in the ratios of the waiting times and not in the absolute values themselves, W_0 can be eliminated from the equations above as follows. We have,

$$\sum_{j=1}^{P} A_{ij} W_j = - W_0\, c \qquad i = 1,2.........P$$

Dividing both sides by W_1,

$$\Rightarrow \sum_{j=1}^{P} A_{ij} (\frac{W_j}{W_1}) = - (\frac{W_0}{W_1})\, c \qquad i = 1,2.........P$$

and it follows that

$$\sum_{j=1}^{P} A_{ij} (\frac{W_j}{W_1}) = \sum_{j=1}^{P} A_{1j} (\frac{W_j}{W_1}) \qquad i = 2.........P$$

$$\Rightarrow \quad \sum_{j=2}^{P} (A_{ij} - A_{1j}) \left(\frac{W_j}{W_1}\right) = A_{11} - A_{i1} \qquad i = 2 \ldots\ldots\ldots P \tag{3}$$

As an added benefit, for the structure of the matrix A given above, the matrix $(A_{ij} - A_{1j})$ is seen to be triangular thereby greatly simplifying the calculation of the waiting time ratios W_j/W_1. In fact, this provides an alternate way to compute the absolute waiting times W_i as follows.

We use an M/G/m approximation to first compute the number of VTs required. We replace the multi-class arrival process by a single process whose arrival rate λ is the sum of the arrival rates of the individual classes. The mean service time x, and the grade of service W, are obtained as the weighted sum of individual service times and grades of service, where the weighting factor for each class is the ratio of its arrival rate to the total arrival rate of all classes. If $\lambda_1,\lambda_2\ldots\ldots\lambda_P$ are the arrival rates of the individual classes and $x_1,x_2\ldots x_P$ their mean service times, the cumulative arrival rate λ of the M/G/m system is given by,

$$\lambda = \sum_{i=1}^{P} \lambda_i \tag{4}$$

the mean service time x by

$$x = \sum_{i=1}^{P} \frac{\lambda_i x_i}{\lambda} \tag{5}$$

and the average Grade of Service W by

$$W = \sum_{i=1}^{P} \frac{\lambda_i W_i}{\lambda} \tag{6}$$

where the W_i's are the specified GOS (waiting times).

The number of VTs is obtained from the M/G/m approximation as the smallest number that will result in a mean waiting time which is not greater than the average grade of service W computed above. To compute this, we use the following well known and accurate approximation for the mean waiting time of an M/G/m system, which is based on M/D/m and M/M/m waiting times.

$$(1-c_v^2)\left(1+\frac{(1-\frac{\lambda x}{K})(K-1)(\sqrt{4+5K}-2)}{16\lambda x}\right)\frac{xC(K,\lambda x)}{K-\lambda x}0.5 + \frac{c_v^2 xC(K,\lambda x)}{K-\lambda x} \tag{7}$$

where c_v^2 is the squared coefficient of variation and $C(k,\lambda x)$ is the Erlang-C formula, with K servers and λx Erlangs of traffic

This approximation was investigated in [5] and [6].

The waiting time ratios are obtained, for a given set of aging factors, from the matrix equations above. Individual *waiting times* are then computed by *partitioning the M/G/m waiting time calculated above among the classes according to their waiting time ratios.*

Computing the absolute waiting times using either equation 1 or the above method involves an approximation -- for equation 1 we need an approximation for W_0 and in the above method we need one for the M/G /m waiting time.

For the two class case [7], the matrix equation (3) above reduce to

$$\left(\frac{c}{\rho_1 + \rho_2}\right)\left(\frac{W_2}{W_1}\right) = \frac{\alpha_1}{\alpha_2} + \left(\frac{c}{\rho_1 + \rho_2}\right) - 1$$

Since the utilization ρ is given by

$$\rho = \frac{\rho_1 + \rho_2}{c}$$

the above can be written as

$$\frac{W_2}{W_1} = 1 - \rho\left(1 - \frac{\alpha_1}{\alpha_2}\right) \quad (8)$$

The reverse problem of computing the aging factors given the waiting time ratio W_2/W_1 is trivial in this case because of the linear relationship between the waiting time ratio and the reciprocal of the ratio of the aging factors for a constant utilization. α_1 is set to 1 and α_2 is computed from the above equation. Encouraged by this, we tried to see if we could apply a similar method for the multi-class case also. We ran a series of simulations wherein the utilization of each class was kept constant. The aging factor of only one class was varied at a time keeping the aging factors of other classes fixed. We saw that for a given class, varying its aging factor, *resulted in a linear variation of its waiting time ratio with respect to the reciprocal of its aging factor*. Moreover, one would expect that changing the aging factor of a particular class would affect the waiting times of other classes, particularly of the lower classes. This was indeed observed, however, *the ratios of the waiting times changed very slightly i.e. though the absolute values of the waiting times of other classes was affected by a change in the aging factor of a particular class, the waiting time ratios seemed to be almost invariant as indicated by the graph below.*

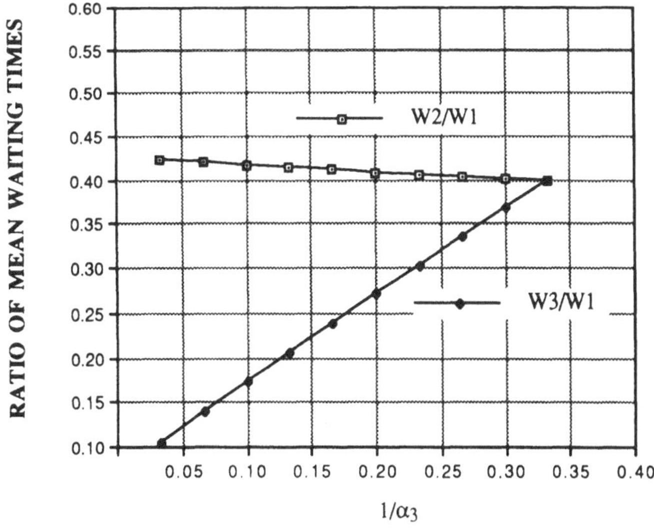

Figure 4: Ratio of mean waiting times as a function of the reciprocal of the aging factor.

The plot above is for a three class case with $\alpha_1 = 1$, $\alpha_2 = 3$, and α_3 is varied keeping the traffic parameters constant. As is evident, the ratio W_3/W_1 varies almost linearly with $1/\alpha_3$ and the ratio W_2/W_1 is practically constant. A large number of numerical computations that were performed indicated that the above result holds independent of the traffic parameters and the number of classes.

This suggests that it might be possible to solve for the set of aging factors that yield the desired waiting time ratios by solving the problem on a class by class basis i.e. we start by fixing the aging factors of all classes except the highest at 1. The aging factor for the highest class is then estimated by a linear approximation as follows. First, for a linear approximation we need two points where the x and y coordinates of a point (x,y) are given by the waiting time ratio obtained and the reciprocal of the aging factor used respectively. The first point is obtained by keeping the aging factor of the highest class also at 1 which naturally yields a waiting time ratio of 1. Hence (1,1) is one of the points. To get the second point, we *arbitrarily* choose another aging factor (a > 1) for the highest priority class and obtain a waiting time ratio "b". (b,1/a) is the second point. Since we have an almost linear relationship of the waiting time ratio with respect to the reciprocal of the aging factor, the reciprocal of the aging factor which yields the *desired waiting time ratio* (W_P/W_1) may be obtained by either interpolating or extrapolating between the above two points. If such a ratio is not achievable, the method typically yields a negative value of the aging factor. We replace negative aging factors so obtained by a pre-determined maximum value (a large number, say 10^6). We fix the aging factor of the highest class at this value and then repeat the same for the lower classes by varying their aging factors between 1 (lower bound) and the aging factor of the next highest class (upper bound) obtained before. During each step of the iteration, the aging factor of *only one class* is varied. At the end of this iteration, we will have our first estimate of the set of aging factors. During successive iterations we modify the upper and lower bounds of each class as follows. At each iteration, we obtain an estimate of the aging factor for a particular class through the above linear approximation technique. With this value for the aging factor we compute the waiting time ratio of that class by solving the matrix equation (3) above. If the waiting time ratio is lower than that desired, we make this estimate of the aging factor the new upper bound since a lower ratio indicates that we have overestimated the aging factor. If the waiting time ratio is higher, we make the estimate of the aging factor the new lower bound since a higher ratio indicates that we have underestimated the aging factor. We continue iterating, each time using a linear approximation between the lower and upper bounds. The test of convergence is to see if all the achieved waiting time ratios are within a previously specified epsilon (typically of the order of 10^{-3}) of the desired ratios. If the desired waiting time ratios are achievable through the use of aging factors, then this approach yields highly accurate results within a few iterations.

In the case where the specified set of waiting time ratios *is not achievable*, the algorithm tries to find an achievable set as follows. We first try to solve for the given set of waiting time ratios. At the end of a specified number of iterations (about 20), we check to see if we have converged. If not, we conclude that the ratios are unachievable. This approach has worked very well, mainly because in those cases where the ratios are achievable, the algorithm normally converges very quickly (within 10 iterations in most cases, and often times within 3 iterations for an epsilon of 10^{-3}). The problem now is to obtain an optimum set of achievable waiting time ratios starting with the given. Defining such an optimum set is very difficult in the general multi-class case as the space of achievable ratios is a very complex multi-dimensional structure.

The approach we adopted to find an achievable set of ratios is to let the algorithm run for a few iterations and then check to see if we are converging as outlined above. If not, we modify the *set of waiting times* as follows. For each class i ($1 < i <= P$), define the *"target" ratio as the waiting time ratio that we want to achieve and the "achieved" ratio as the waiting time ratio that was actually achieved* by the algorithm at the end of a *specified* number of iterations. For the highest class we choose as the new target ratio the *average of its previous target and achieved ratios*. The *waiting time of the lowest* class(class 1) is then modified to reflect the new target ratio for the highest class by scaling it down appropriately. The waiting times of successive higher classes are modified *only if their current waiting times happen to be higher than the modified waiting time of the next lower class.* The new waiting times of each of the successive higher classes that are modified are obtained by *scaling down their previous waiting times in the same ratio that the lowest class's waiting time was scaled down.* For instance, in a "P" class case, let $W_1, W_2, W_3 \ldots\ldots W_P$ be an unachievable set of desired waiting times. Let the target set of waiting time ratios $(W_2/W_1, W_3/W_1 \ldots W_P/W_1)$ be denoted by $t_2, t_3, \ldots t_P$. Let $a_2, a_3 \ldots a_P$ be the set of waiting time ratios that were actually achieved by the algorithm in the first iteration. For the second iteration, we obtain a new set of waiting times as follows. The new target ratio of the highest class, t_P, is given by

$$t_{P\,(new)} = (\, t_{P\,(previous)} + a_{P\,(previous)})/2.$$

The waiting time of the lowest class W_1 is then given by

$$W_{1\,(new)} = (W_{1\,(previous)} * t_{P\,(previous)})/t_{P\,(new)}$$

If $W_{i\,(previous)} > W_{i-1\,(new)}$, then it is modified as

$$W_{i\,(new)} = W_{i\,(previous)} * (W_{1\,(new)}/W_{1\,(previous)})$$

From the modified set of waiting times, the new target set of waiting time ratios $(W_2/W_1, W_3/W_1 \ldots W_P/W_1)$ is computed. We repeat this procedure until we find a set of achievable waiting time ratios.

Ideally, given a set of ratios that lie outside the achievable space, we would like our algorithm to find a point on the boundary of the achievable performance space closest to the specified one. As stated before, it is very difficult to visualize this optimum point in the general multi-class case. Given such an unachievable point, the approach presented above may yield a point within the achievable space, instead of a boundary point. As an improvement, one could now use a bisection method between the set of ratios that were achieved and the most recent set of ratios that could not be achieved during the iterative process. The new target ratio of the highest class is now chosen as the midpoint of the *most recent "achieved" and the "non-achieved" ratios.* The waiting times and the target ratios of the lower classes are obtained as before. One would expect this improvement to give us a point closer to the optimum boundary point, if not the optimum point itself.

V. NUMERICAL RESULTS

The following tables illustrate the algorithm for provisioning VTs and determine the appropriate aging factors -- α_i. λ and x are the arrival rates and the mean service times (for the simulation run, the service times were assumed to be exponential) respectively, GOS(W) is the specified Grade of Service in terms of the mean waiting times, W is the actual waiting time achieved using the aging factors α that were generated by the algorithm. GOS(w) is the user specified set of ratios (obtained from GOS(W)) and w is the actual set of ratios (using equation 3) that were achieved using the computed the aging factors. The "Simulation" column is the set of waiting time ratios that are computed via a discrete event simulation of $30*10^6$ arrivals -- requests for access to the network. With the exception of the simulation column, the text in italics is program output and the normal text is program input.

TABLE 2: AGING FACTORS FOR A FEASIBLE SET OF WAITING TIME RATIOS

CALLQ	λ	x	GOS(W)	W	α	GOS(w)	w	Simulation
1	0.1	10	10	*7.02*	*1*			
2	0.2	11	9	*6.32*	*1.12*	0.900	*0.900*	0.906
3	0.3	12	8	*5.61*	*1.28*	0.800	*0.800*	0.802
4	0.4	13	7	*4.91*	*1.48*	0.700	*0.700*	0.697
5	0.5	14	6	*4.21*	*1.74*	0.600	*0.600*	0.599
6	0.6	15	5	*3.51*	*2.12*	0.500	*0.500*	0.499
7	0.7	16	4	*2.81*	*2.69*	0.400	*0.400*	0.397
8	0.8	17	3	*2.11*	*3.66*	0.300	*0.300*	0.297
9	0.9	18	2	*1.40*	*5.69*	0.200	*0.200*	0.198
c	:	*72*						
ρ	:	*0.9583*						

In table 2, the mean Grade of Service for the M/G/m approximation is computed from the specified GOS for the individual classes (GOS(W)) using equation (6) as 4.67, the total arrival rate is 4.5 (the sum of the individual λ) and the mean service time is 15.33 (from equation (5)). The minimum number of servers that will lead to a mean waiting time (as computed in (7)) that is less than 4.67 is 72; with 71 servers the mean waiting time is 5.77 and with 72 it is 3.28. For the given GOS(W), we were able to exactly achieve the waiting time ratios as indicated by the identical numbers in column GOS(w) and w, however, the specified absolute values of the waiting times are not achieved. The M/G/m mean waiting time W = 3.28, is now partitioned among the individual classes (W_i) according to the ratios achieved (column w) using the aging factors above (in column α). Note, the mean service times (x_i) are not identical, thus equation (1) is an approximation, and consequently the method for calculating aging factors is also an approximation. However, the simulation results indicated the approximation worked well.

TABLE 3: AGING FACTORS FOR AN INFEASIBLE SET OF WAITING TIME RATIOS

CALLQ	λ	x	GOS(W)	W	α	GOS(R)	w	Simulation
1	0.01	10	15	*7.92*	*1*			
2	0.02	11	9	*5.11*	*3.29*	0.60	*0.64*	0.634
3	0.03	12	8	*4.54*	*4.12*	0.53	*0.57*	0.561
4	0.04	13	7	*3.97*	*5.23*	0.46	*0.50*	0.488
5	0.05	14	6	*3.40*	*6.79*	0.40	*0.43*	0.415
6	0.06	15	5	*2.84*	*9.17*	0.33	*0.36*	0.344
7	0.07	16	4	*2.27*	*13.20*	0.27	*0.29*	0.274
8	0.08	17	3	*1.70*	*21.45*	0.20	*0.21*	0.205
9	0.09	18	2	*1.14*	*46.33*	0.13	*0.14*	0.137
c	:	*9*						
ρ	:	*0.7667*						

The parameters of the M/G/m approximation are calculated as in table 2. However, the ratios specified are not achievable in this case. Specifically, it is not possible to achieve the ratio, w_9, of 0.13 (2/15) between class 9 and 1. The algorithm then tries to achieve the lowest possible ratio (> 0.13) according to the method outlined in the analysis section. This ratio is seen to be 0.14 for an aging factor of 46.33 for class 9. It is interesting to note that the ratios between the other classes are maintained. For example, from the GOS(W) column we see that the desired ratio between class 9 and class 8 are maintained since the desired ratio was 2/3 = 0.67 and the achieved ratio was 1.14/1.70 = 0.67. In fact, the only ratios that were not achieved are those defined with respect to the lowest priority class -- class 1.

VI. SUMMARY

In this paper we provide accurate analytical approximations to compute both the aging factors and the number of Virtual Tributaries (VTs) required to yield a Grade-of-Service (GOS), where the GOS is assumed to be specified in terms of expected waiting time for each access queue. The number of Virtual Tributaries is obtained using an M/G/m approximation. Since this number has to be an integer, the absolute values of the waiting times may not be achieved exactly. However, using an iterative approach we were able to determine aging factors that would yield the desired waiting time ratios whenever such a set of waiting time ratios was feasible.

REFERENCES

[1] John Bellamy, "Digital Telephony", Second Edition, New York, Wiley, 1991.

[2] L. Kleinrock, "A Delay Dependent Queue Discipline", Naval Research Logistics Quarterly,Vol. 11, pp 329-341, 1964.

[3] L. Kleinrock, "Queueing Systems Vol. II: Computer Applications", New York: Wiley, 1976.

[4] A. Federgruen and H. Groenevelt, "M/G/c Queueing Systems with Multiple Customer Classes: Characterization and Control of Achievable Performance Under Nonpreemptive Priority Rules", Management Science, pp 1121-1138, Vol. 34, No. 9, Sept. 1988.

[5] G.P. Cosmetatos, "Some Approximate Equilibrium Results for the Multiserver Queue M/G/c", Operations Research Quarterly, vol. 27, pp 615-620,1976.

[6] E. Page, "Queuing Theory in OR", Butterworths London, 1972.

[7] M. Perry and A. Nilsson, “Performance Modeling of Automatic Call Distributors: Assignable Grade of Service Staffing”, XIV International Switching Symposium, October 1992, Yokohama, Japan.

A REVIEW OF VIDEO SOURCES IN ATM NETWORKS

Philip F. Chimento Jr.
E95/B673
IBM Corporation
Research Triangle Park, N.C. 27709

Abstract

In this brief paper, we discuss the transport of digital video data in the Broadband ISDN environment. We review the characteristics of video traffic streams that are generated by coder/decoder equipment and talk about some attempts to characterize statistically the source stream of video cells.

Introduction

The promise of Broadband ISDN is a network providing integrated services to its customers, including voice, data, image and video. The mixture of all these services on a single network fabric is a challenge in a number of areas, such as: service interfaces, resource allocation, structure of switching fabrics, QoS guarantees, not to mention charging, accounting and maintenance.

The economic incentive for providing such services is great. Providing businesses with high quality videoconferencing facilities is just one potential application. Another is providing instructional film, on demand, to remote locations, for example, supplementing repair manuals with videos. Furthermore, entertainment applications of video distribution, such as video on demand, hold tremendous potential revenue.

In this paper, we want to look at some of the requirements that a range of video services places on a network. In the Integrated Video Services Baseline Document [3], CCITT SG XVIII outlines the requirements for video services, and at the same time raises a lot of questions about how to provide those services. In general, the CCITT describes four types of video services that an ATM network is able to provide: First, there are distribution services. These are essentially broadcast services, where a source may distribute a video signal to many users. In this case, communication is generally unidirectional. Second, there are conversational services. Examples of this type of service are videotelephony and videoconferencing. In this case, information may flow from many sources to many destinations. Third, there are messaging services. An example of this is the moving picture analogue to voice mail. Finally, there are retrieval services, where images or video sequences are retrieved from libraries. These last two

Asynchronous Transfer Mode Networks, Edited by Y. Viniotis
and R.O. Onvural, Plenum Press, New York, 1993

types of services may have multiple destinations also, or they may be primarily point-to-point.

These service offerings have broad implications on the functions that an ATM network must provide. For example, distribution and conversational services require point-to-multipoint and multipoint-to-multipoint call setup and maintenance procedures, in addition to such services as adding a party after the call has begun or dropping a part before the call is over. But perhaps the most difficult set of problems to be solved, particularly for video, are those having to do with service guarantees, traffic contracts with the users, traffic descriptors, network resource allocation, and network traffic control.

There are two benefits expected from using Variable Bit Rate (VBR) video instead of Constant Bit Rate (CBR) video. On the user side, the benefit is a constant quality video signal delivered at the output device. From the network perspective, the major potential benefit is the ability to gain in the number of video connections multiplexed on the network facilities. What makes it difficult to obtain these benefits is the nature of the video traffic stream and its service requirements. Video is extremely high bit rate, compared to both single voice and data sources. Uncompressed television-quality video can require upwards of 300,000 samples per frame and requires 30 frames per second [15]. This translates into raw bit requirements of 7 to 9 Megabits per second. Of course the raw stream will be compressed, but still the bit rate requirements are high. In addition, the delay requirements for video sources are tight. Frame rates of 25 or 30 frames per second dictate that the end-to-end delay for a VC or VP carrying video traffic be within the 30-40ms time limit (minus the time required to decode, uncompress and play out the stream at the receiving end). Otherwise, large playout buffers at the receiver may be required. Another requirement is that in order to maintain picture quality, cell loss must meet stringent requirements. In [3], the requirements for uncompensated cell loss ratio for video are on the order of 10^{-7} to 10^{-10}. In [8], the requirement is for a cell loss ratio of 10^{-11}. In contrast, data sources, which generally have a retransmission protocol above the ATM and AAL layers, can tolerate much higher cell loss ratios. Retransmission of video cells seems impractical because of the time constraints and so to avoid a degradation of quality, effective cell loss ratios must be small. This requirement can be alleviated somewhat by block coding techniques to compensate for the loss of a cell, but this introduces additional complexity into the playout equipment. Finally, VBR video is highly bursty. As we shall see later, scene changes and motion force many more bits to be generated per frame than a relatively quiet "talking head".

So, on the one hand, we have fairly stringent service requirements to be met. On the other hand, the ATM network needs to allocate resources efficiently and protect itself and other users against misbehaving sources and some high stress events such as many sources bursting simultaneously. The ATM network generally has two mechanisms it can use to achieve these goals: Usage Parameter Control (UPC) and intermediate node cell handling.

UPC generally takes the form of a "leaky bucket" mechanism. See, for example [4]. The leaky bucket is an algorithm that admits source cells at a maximum negotiated rate. This mechanism in effect polices the sources by not allowing a given source to exceed the contract implied by the traffic descriptors the source gives to the network. In the case of video, however, the traffic descriptors may be complex. This is because some cells generated by a video codec will carry information that is more important to picture quality than others. This naturally divides the video source into two streams or layers, one with relaxed cell loss ratio requirements and the other will stringent requirements. One challenge is to devise UPC mechanisms for handling these two streams to achieve both inflow control and cell loss ratio control.

There are several issues in intermediate node cell handling. First, we should note that the buffers at VP or VP/VC switching points in the ATM network must be relatively small or else the end-to-end delay and delay variation may be unacceptably large. Therefore, during transient overload conditions, the network may be obliged to discard cells. There is a danger that with periodic or near-periodic sources merged on the same transmission facility, some source may see discards in bursts or periodic degradation of service [14], [5].

It should be apparent that the study of video source characteristics is crucial for addressing these questions. Researchers are active in all of these areas. The bibliography of this paper contains only a small fraction of the work that has been done on this topic in the past couple of years. In the remainder of this paper, we will go into a little more detail on the structure of video streams and then discuss briefly a few of the many models that have been proposed.

The Coding of Video Signals

In this section, we will look at the generation of the digital information streams that represent the analog video signals. As noted in the introduction, current video technology has a large bandwidth requirement that would be expensive to transmit uncompressed. In addition, the raw bandwidth requirements of new technologies such as HDTV promise to be even greater. Over the past few years, much work has been published on various coded designs, exploring different ways of reducing the bandwidth requirements and maintaining picture quality. Some examples are given in [9, 10, 6, 7, 13]. This list is by no means exhaustive ! Instead of an extensive survey, we will look at the MPEG standard.

Here we summarize the MPEG coding algorithm as described in [17] and [1]. MPEG is the standards group in ISO (ISO-IEC/JTC1/SC2/WG11) that is addressing the compression and synchronization of video and audio signals so that the video stream will be about 1.5 Mbits/second. This would allow the video stream to fit on a T1 (1.536 Mbits/second) transmission facility. There are other coding standards (e.g. H.261) aimed at other design points. The MPEG standard was designed to meet the requirements of different applications and so is paramterized to meet application needs. The standards committee has, however, designed a core parameter set that meets the basic design points and which all decoders should be able to handle.

Some key requirements considered by the MPEG group that are of interest to the networking community are: Random access of frames (target access time is 500 milliseconds); coding and decoding delay (target for videotelephony is 150 milliseconds, for other applications, 1 second). The first requirement sets delay limits on the network round trip time, while the second points out that the total application-to-application delay budget may have significant portions allocated to application layer functions.

Video sequences have two types of redundancy that can be exploited: spatial and temporal. The temporal sequence of frames in MPEG consists of three kinds of frames: "I"-pictures or intra-pictures; "P"-pictures or predicted pictures; "B"-pictures or bidirectionally interpolated pictures. These three picture types are an attempt to strike a balance between the random access of pictures and compression levels. I-frames provide reference points for random access functions, but do not provide as much compression as other frame types. B-frames, because they are interpolated with respect to both "future" and "past" reference frames provide the highest compression ratios. P-frames are coded using motion-compensation prediction techniques with reference to either I-frames or other P-frames. The interleaving of these frame types in a video sequence may be application dependent. As noted in [17], this may produce interesting correlations in the stream as seen by the telecommunications system. The network is concerned with the number of cells produced per frame, which forms the arrival stream for a VC or VP providing the network service. Depending on the picture content and motion, a frame sequence such as IBBBPBBBPBBBI... could produce a large number of cells for the I-frames and P-frames and smaller numbers of cells for the B-frames. The autocorrelation function of the number of cells per frame might then show large correlations at lags that are multiples of 4 and small or negative correlations at lags that are multiples of 1, 2, and 3. (This is partially borne out by evidence in [1].)

For purposes of coding motion compensation, the frame is divided into 16x16 macroblocks. Motion prediction is done on the basis of translated motion (but not rotational [15]). Motion vectors are associated with each macroblock and are represented using differential coding with respect to adjacent blocks. The bit requirements are further reduced by using a variable-length code on the motion vectors. The motion vectors themselves are computed using a block-matching algorithm that minimizes an implementation-determined cost of block mismatch. In addition, prediction error resulting from motion compensation is coded using the Discrete Cosine Transform (DCT) and variable-length codes.

Since motion compensation is based on macroblocks, MPEG tries to reduce spatial redundancy also based on blocks (8x8 blocks, 4 per macroblock). MPEG uses the DCT for this compression block by block. Each block of image pixels is converted to the frequency domain by the DCT, producing a block of 64 transform coefficients. High frequency components, such as those produced by the differential coding of P-frames and B-frames are quantized coarsely; I-frames contain weight in many frequencies and need to be quantized on a finer grid. At the output, the DCT coefficients are compressed with a variable-length code.

The data stream produced by an MPEG coder contains different types of information which may be produced at different rates, depending on how the coder is designed. The detailed behavior of these information streams can have an effect on the UPC mechanism and on resource allocation in the network. As pointed out in [1], relatively little work has been done on these detailed characterizations. In [1] they characterize the MPEG output stream by the distribution of cells per frame and by the distribution of burst interarrival times for different quantization levels (bit rates) and different mixes of I-frames and P-frames. (They ignore B-frames for their study.) In [13], they use a different codec compression scheme and determine the empirical distributions of packet interarrival times for different classes of information output from the codec: Prediction errors and motion vectors. In both these papers, the authors found that the distributions in question were complex distributions–perhaps mixtures of different distributions.

The detailed behavior of the different information streams produced by a codec may affect the traffic descriptors that an ATM network requires. In the examples just discussed, some MPEG (or other) codec output may be classified as Cell Loss Priority (CLP)=0 (important information) while other information may be classified as CLP=1 (less important). If (as evidenced by [13]) the inter-packet generation interval distributions for CLP=0 and CLP=1 traffic streams are very different, then it may force these streams to be monitored and policed separately. This is because the a burst from the merged stream may cause important information to be thrown away as violating the traffic contract, even though the CLP=0 stream is within its limits. Further, the characterization burden on the network user is increased, because he or she must characterize the sub-streams individually in order to have an accurate traffic contract with the network.

Another interesting point is made in [16] that again, depending on codec design, there are different possibilities for cell generation: (1) The codec could generate cells at the peak rate of the connection between the codec and the network in a single burst. The durtion of the burst is less or equal to a frame time. (2) The codec could generate cells continuously at a constant rate. The rate is determined by the total number of cells for the frame divided by the frame time. (3) The codec could generate cells as blocks are coded causing perhaps several bursts within a frame time. (This mode was assumed by [1, 13].) Each of these cell generation modes will affect both the UPC algorithm operation and resource allocation.

Models of Video Sources

In this section, we discuss very briefly several models of video sources found in the literature that are based on empirical data. Note that there are Markovian models (for example MMPPs) that are also used to model video (and other) sources. However, we do not cover those types of models in this paper.

ARMA Models

Autoregressive-Moving Average (ARMA) models are used to model empirical time series. That makes them a natural choice to model sequences from video codecs. In this section, we report briefly on 3 ARMA models found in the literature in the last few years. There seems to be no consistency of findings.

An ARMA model has the form

$$x_t = \sum_{i=1}^{p} \alpha_i x_{t-i} - \sum_{j=1}^{q} \beta_j w_{t-j} + w_t$$

where the $\{w_t\}$ process is uncorrelated white noise and the $\{\alpha_i\}$ and $\{\beta_j\}$ are estimated from the time-series realization.

Model 1

In [6], the authors analyze two videoconferencing scenes, one active and one inactive. The frame rate is 15 frames/second and each scene lasts 20 seconds. The scenes are steady: no panning, zooming or scene changes take place. The authors fit AR(1) models to the time series of bits generated per frame for each scene. They also fit Gaussian models to the normalized bit rate distributions. The AR(1) model tracks the empirical autocorrelation function for about the first 6 lags in each scene, and then it diverges. The Gaussian models of the bit rate distributions tend to be a poor visual fit, but the authors do not provide any 'goodness of fit' statistics.

The authors use their source model to feed an infinite buffer, single server (fixed rate) queue to try to model superimposed video sources. In determining the delay distributions from the model, their approximate source model seems to do well compared to the simulations driven from the original data.

Model 2

In [12], a much more elaborate analysis is performed. In that paper, the authors consider the number of video calls arriving within a fixed time slot (about 123 μseconds) as the time series. There were three video sequences producing the three time series that they analyzed. Each sequence was 4 seconds long, and the frames were transmitted at 25 frames/second. The scene content ranged from videoconferencing to a low flying airplane. They found that the sample autocorrelation function of each sequence exhibited strong correlations at 1 frame intervals. Their explanation for this is that there is a strong frame-to-frame correlation of information. Since their data was so fine-grained, they had to fit a rather large ARMA model (order 325) to capture the correlations observed in the sample autocorrelation functions.

They used their source models as input to a G/D/1 model of a multiplexer and simulated the system. They compared the correlated (ARMA) source models to uncorrelated source models and found through simulation that the waiting time at heavy load was greatly affected by the correlated sources.

Model 3

In [14], the authors considered a sequence generated by a 30 minute videoconference, with relatively little motion and no scene changes panning or zooming. The frame rate was 25 frames/second. They fit a gamma distribution to the empirical number of cells/frame distribution and used a Q-Q goodness of fit test. They tried fitting AR(1), AR(2) and AR(3) models to the empirical time series. The AR(1) model did not have as good a fit statistically as the AR(2) model did. The AR(3) model was found to have the α_3 coefficient significantly different from zero. So the AR(2) model seemed to have the best fit. The authors also checked the empirical series for stationarity and it was found to be stationary.

The authors used the AR(2) model to drive a simulation of a multiple source finite buffer multiplexer. They found that the AR(2) model did not generate enough frames with a large number of cells and so underestimated the cell loss probability compared to simulations driven with the original data sequences. Given that, they turned to a detailed Markov model of the frame generation process and that seemed to work satisfactorily.

Histogram Model

The only model in this category is given in [2]. The approach that the authors take is to collect source bit rate histograms (bits generated per frame) from 10 second video sequences. The histograms are only 8 bins wide and so are compact. These histograms do not capture the time-dependent behavior of the sources and so do not model correlations within a data stream.

The authors model a multiplexer queue as an M/D/1/N system and their objective is to compute steady state buffer occupancy distributions. To use their histogram source models, they make a number of assumptions: First that cell generation times are uniformly distributed over a frame interval. Second, that the system reaches steady state in an interval that is small compared to a frame interval.

If we admit these assumptions, then letting λ_i be the cell arrival rate with probability p_i, and letting $P(n)$ be the probability that there are n in the M/D/1/N queue (approximated), then we have $P(n) = \sum_i P(n|\lambda_i)p_i$. They tested their model against simulations of the system and found reasonable agreement.

Comments and Conclusions

We found in general no agreement among the researchers using empirical based models of video sources. Part of the explanation seems to lie in the fact that the scene content, codecs and length of the time series vary a lot. The stationarity found by [6, 14] and others may have a lot to do with the relative stability of the scenes, but it may also depend on the granularity of the data that they used. The results from [12] hint that that may be the case.

In any case, [14] identified the "source periodicity" effect from their investigations which is an important phenomenon. The "source periodicity" effect is the phenomenon of certain multiplexed sources seeing cells consistently dropped and consequently experiencing much larger cell loss rates than other sources multiplexed on the same facility. This occurs due to the periodicity of the sources and in particular for video sources since they have a fixed period of one frame time.

An attempt to avoid the dependence of models on scene content and coding algorithm can be found in [11]. In that paper, the authors propose a set of indices measuring the spatial and temporal correlation of the sequence of frames. They then try to characterize a coding algorithm as a linear combination of the proposed indices.

We have reviewed some of the requirements for the transport of video traffic in B-ISDN networks, taken a look at the MPEG coding algorithm and some possible effects on traffic control in ATM. Lastly, we have briefly reviewed some video traffic models based on video data. In future reports, we hope to report some more detailed investigations of our own on these topics.

References

[1] P. Pancha, M. El Zarki, "A look at the MPEG video coding standard for variable bit rate video transmission" *INFOCOM '92,*, pp.1C.2.1-1C.2.10.

[2] Paul Skelly, Sudhir Dixit, Mischa Schwartz, "A Histogram based Model for Video Traffic Behavior in an ATM Network Node with an Application to Congestion Control", *INFOCOM '92,*, pp.1C.3.1-1C.3.10.

[3] CCITT Study Group XVIII IVS Baseline Document, Geneva, June 1992.

[4] CCITT Study Group XVIII Recommendation I.371, Geneva, June 1992.

[5] V. Ramaswami and W. Willinger, "Efficient traffic performance strategies for packet multiplexers", *Computer Networks and ISDN Systems* Vol. 20, No. 1-5, pp.401-408.

[6] M. Nomura, T. Fujii, N. Ohta, "Basic Characteristics of Variable Rate Video Coding in an ATM Environment", *IEEE Journal of Selected Areas in Communications*, Vol. 7, No. 5, June 1989, pp.752-760.

[7] Y. Zhang, W.W. Wu, K.S. Kim, R.L. Pickholtz, J. Ramasastry, "Variable Bit-Rate Video Transmission in the Broadband ISDN Environment", Proceedings of the IEEE, Vol. 79, No. 2, February 1991, pp.214-221.

[8] W. Verbiest, L. Pinnoo, B. Voeten, "The Impact of the ATM Concept on Video Coding", *IEEE Journal on Selected Areas in Communications* Vol. 6, No. 9, December 1988, pp.1623-1632.

[9] G. Karlsson, M. Vetterli, "Packet Video and Its Integration into the Network Architecture", *IEEE Journal on Selected Areas in Communications* Vol. 7, No. 5, June 1989, pp.739-751.

[10] W. Verbiest, L. Pinnoo, "A Variable Bit Rate Video Codec for Asynchronous Transfer Mode Networks", *IEEE Journal on Selected Areas in Communications* Vol. 7, No. 5, June 1989, pp.761-770.

[11] R. M. Rodriguez-Dagnino, M.R.K. Khansari, A. Leon-Garcia, "Prediction of Bit Rate Sequences of Encoded Video Signals", *IEEE Journal on Selected Areas in Communications* Vol. 9, No. 3, April 1991, pp.305-314.

[12] R. Grunenfelder, J.P. Cosmas, S. Manthorpe, A. Odinma-Okafor, "Characterization of Video Codecs as Autoregressive Moving Average Processes and Related Queueing Systems Performance", *IEEE Journal on Selected Areas in Communications* Vol. 9, No. 3, April 1991, pp.284-293.

[13] R. Kishimoto, Y. Ogata, F Inumaru, "Generation Interval Distribution Characteristics of Packetized Variable Rate Video Coding Data Streams in an ATM Network", *IEEE Journal on Selected Areas in Communications* Vol. 7, No. 5, June 1989, pp.833-841.

[14] D. Heyman, A. Tabatabai, T. Lakshman, "Statistical analysis and simulation study of viedo teleconference traffic in ATM networks", *IEEE Transactions on Circuits, Systems and Video Technology* Vol. 2, No. 1, March 1992, pp.49-59.

[15] T. Mitchell, S. Blake, "Techniques for Video Transport at the ISDN Basic Rate", *N.C. State University, Center for Communications and Signal Processing* Technical Report TR-92/8, June 1992.

[16] J. Roberts, J. Guibert, A. Simonian "Network performance considerations in the design of a VBR codec" in "Queueing, Performance and Control in ATM", Proceedings of the Thirteenth International Teletraffic Congress, ed. Cohen and Pack, June 1991, North Holland.

[17] D. LeGall, "MPEG: A video compression standard for multimedia applications", *Communications of the ACM*, Vol. 34, No. 4, April 1991, pp 47-58.

MULTIMEDIA NETWORKING PERFORMANCE REQUIREMENTS

James D. Russell

IBM Networking Systems
P.O. Box 12195, Dept. H40/656
Research Triangle Park, North Carolina 27709

I. THROUGHPUT DEMAND IN MULTIMEDIA APPLICATIONS

"Throughput Demand" (bits per second per workstation) is a measure of average communications traffic placed on a networking system per user in particular applications. This has been defined for several applications in an earlier report (5).

In estimating Throughput Demand for multimedia applications, it is first necessary to understand the particular combinations of information types (alphanumeric, graphics, image, audio, video) to be communicated. For example, voice annotated text as might be communicated to support an education lecture involves only two forms of information (audio and alphanumeric) each of which are low in Throughput Demand. At a higher extreme, a "Complex Teleconference" can concurrently utilize all information types (alphanumeric, graphics, image, audio, and video) and require significant bandwidth and complicated synchronizations. An example given in (1) describes a medical teleconference involving audio and video showing two doctors discussing diagnostic information presented in graphics, image and text form.

Figure 1 summarizes the types of information communicated in multimedia applications and provides typical figures on object sizes and data rates. The total Throughput Demand of a multimedia application can be estimated by understanding the combination of information types involved, the rate at which the discrete information is updated (e.g., how often does the user see an updated image or an updated graph) and the audio and video data rates used in the application. We will estimate Throughput Demand for some typical multimedia applications below, but first we consider the factors which influence audio and video data rates.

Figure 2 summarizes factors which determine video and audio data rates and Figure 3 identifies data rates typical of several applications. As shown, there is significant variation depending on the users audible and visual quality requirements, and the

Asynchronous Transfer Mode Networks, Edited by Y. Viniotis
and R.O. Onvural, Plenum Press, New York, 1993

Type of Information	Discrete units of Data -- Typical Size:
● **Alphanumeric coded data** Text and numeric data coded as in EBCIDIC or ACSCII code	1240 Bytes per page (IBM 3270 transaction)
● **Graphics** Vector end points and geometric orders.	35,000 Bytes per page (IBM 6090 CAD/CAM)
● **Image** "bit map" or "raster pattern".	60,000 Bytes per page (Office Image) 600,000 Bytes per page (medical, manufacturing)

Type of Information	Continuous Data Streams -- Typical rates:
● **Audio** 15Hz - 20KHz sound, compressed.	2.4 to 384 Kbits/sec - depending on compression ratio and required audible quality.
● **Video** moving pictures	64 to 4000 Kbits/sec - depending on compression ratio and required visual quality.

Figure 1 - Media Types

degree of information compression which can be achieved. With sufficient compression and with tolerance for low quality sound, audio information can be communicated with as little as 2400 bits per second of bandwidth. In this case, voices are a barely intelligible robotic sound and the identity of the speaker cannot be determined. At the other extreme, 384,000 bits per second can produce CD quality stereo sound. As shown, similar trade-offs in video information will influence the data rate requirements. Depending on the compression method, low data rate video will often result in blur, blocking, jerky motion and other artifacts (6). Higher data rates eliminate these effects and can provide crisp colorful images with smooth motion comparable to the output of a video cassette recorder.

With the understanding that multimedia Throughput Demand depends on the application variables identified above, we can identify some hypothetical cases for the

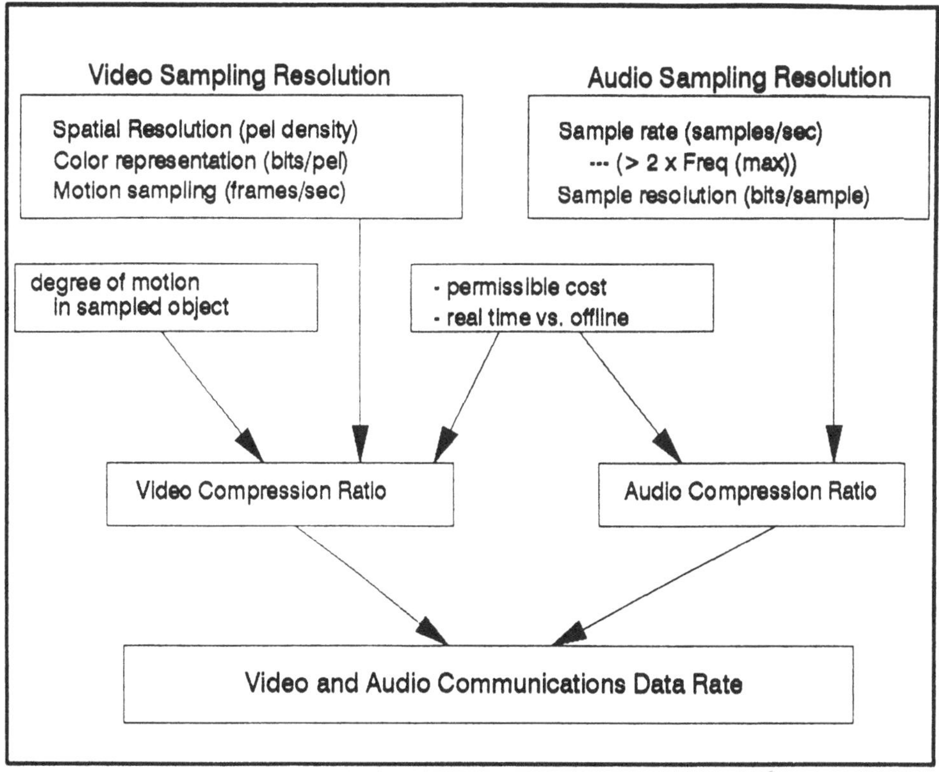

Figure 2 - Video and Audio Data Rate Determinants

purpose of evaluating communications performance requirements. These cases are shown in Figure 4.

The present conclusion is that Throughput Demand of typical near term multimedia applications ranges from about 32,000 bits per second for telephone quality voice-annotated-text applications as in an education lecture to as high as 1.5 million bits per second with high quality sound and video as typical of advertising or home video distribution.

Dramatic improvements in video compression capabilities are expected before the mid 1990s (CCITT H.261). This could lead to reduction in video data rate requirements. However, higher quality influences such as High Definition TV may lead to overall growth in throughput required of future multimedia communication systems, depending on implementation costs. Moreover, future multimedia Throughput Demand will depend on future user quality requirements, improvements in audio and video compression performance, and achievable costs. Hence, the quantitative analysis given above is representative of applications during the early 1990s and may vary in future years.

	Information Type	Kbits/sec/session (average)	Notes
Audio	Minimal Intelligible voice	2.4	LPC Coding (robotic sound)
	telephone quality- FM radio quality- Stereo CD Quality-	32 132 384	TI Digital Signal Processor (Byte Magazine, 5/89 Pg.21)
VIDEO-Real Time	128 x 120 pels 15 frames/sec 9 bits color	288	Intel DVI Real Time Video (typical figure at four Delta frames per Reference frame).
	128 x 120 x 15 fps 256 x 240 x 15 fps 256 x 240 x 30 fps	64 384 2000	CCITT H.261 Standard (mid - 1990s)
VIDEO - Non Real Time	Low Resolution Server 352 x 240 pels 10 frames per sec. grayscale,low res color	384	Data Communications, 9/91, page 33. CLI Rembrandt II Picturetel 20 MCI Prism Service
	VCR Quality Server 352 x 240 pels 30 frames per sec. 24 bit color	1100	Intel DVI Presentation Level Video Communications of the ACM, July 1989, 32-7-page 811
	Studio QualityServer 640 x 480 pels 30 frames per sec 24 bit color	4000	MPEG - Thompson Co. Proprietary - UVC Inc.
Future	VIDEO -- HDTV 1125 lines 30 fps 24 bit color	Selectable: 60,000 97,000 120,000	Canon Inc, 9/91 announcement in Japan (uncompressed data rate = 800 Mbps)

Figure 3 - Audio & Video Data Rates (compressed)

II. A VARIETY OF APPLICATIONS AND A VARIETY OF REQUIREMENTS

It is important to recognize different types of multimedia applications because system performance requirements are distinguished among these application types. One distinction is between conversational or interactive applications where delays in communication, or "latency", are critical to the natural flow of information from one person to another, as compared to a distributive or one-way communication where the delay in initiating the communication is not as critical. For distributive applications such as fetching a film clip from an audio/video server, the user will

Multimedia Application Examples	Throughput Demand per active session
Voice annotated text -- education lecture, telephone quality voice one page form per 30 secs. of voice. (1240x8 bits)/30sec. + 32000 bps	bits/sec 32,331
Voice annotated Office Image -- Insurance Claims Processing Insurance Form and 90 sec interview (60,000x8 bits/90sec) + 32000 bps	37,333
Voice Annotated Hi Res Image -- Medical Diagnosis XRAY with 90 sec voice comments $\frac{2000x2000x12 \text{ bits}}{10 \text{ compression}} \times \frac{1}{90 \text{ sec}}$ + 32000 bps	85,333
Basic Teleconference -- AT&T/NCR Teller Terminal Prototype Communications Week, 11/25/91, Page 18	128,000
CD Quality Sound & Office Image -- Library Systems Music score and 60 sec sound (60,000x8 bits/ 60sec) + 384000 bps	392,000
Complex Teleconference -- Real time audio video conference with graphics, image, and text references Little et al, IEEE Network Magazine, 11/90	650,000
Video Distribution System -- Video Distribution System CD Quality sound & VCR Quality Video 1100Kbps + 384 Kbps	1,484,000

Figure 4 - Multimedia Throughput Demand Examples

tolerate seconds of delay in starting the readout, but for a conversational application, delays of more than a few hundred milliseconds cause a loss of natural reactions in the two-way conversation.

In the examples shown in Figure 4, "Basic Teleconference" and "Complex Teleconference" are conversational applications which have more demanding latency objectives than the other examples which are distributive applications.

- <u>LATENCY</u> : the average delay between sampling and presenting the start of the data stream representing a block of information. Latency includes sample,encode,packetize, transmit, decode, and presentation delays.

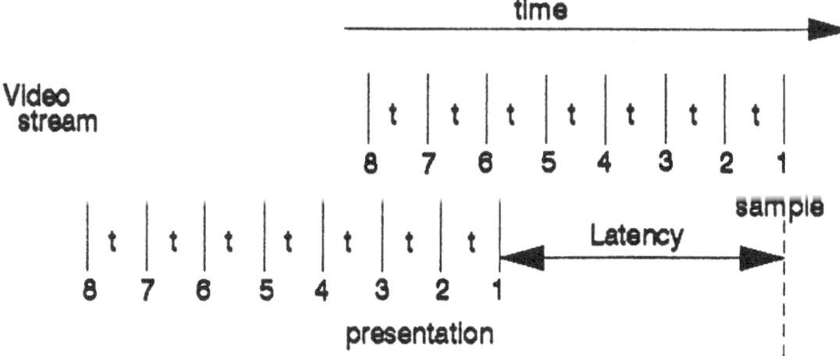

- <u>JITTER</u> : an instantaneous variation in object presentation time.

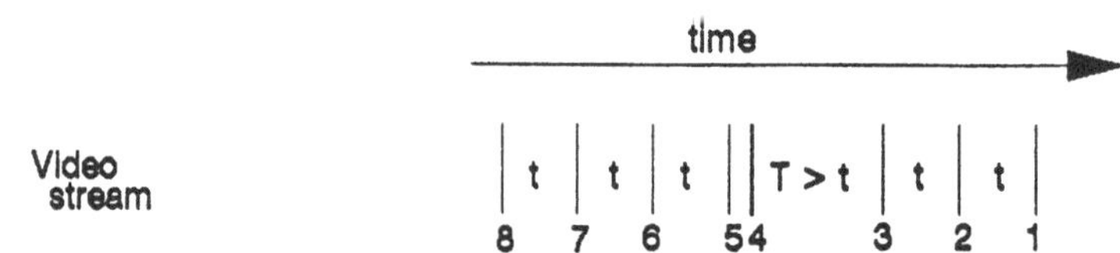

- <u>SKEW</u> :difference in presentation times between two related objects.

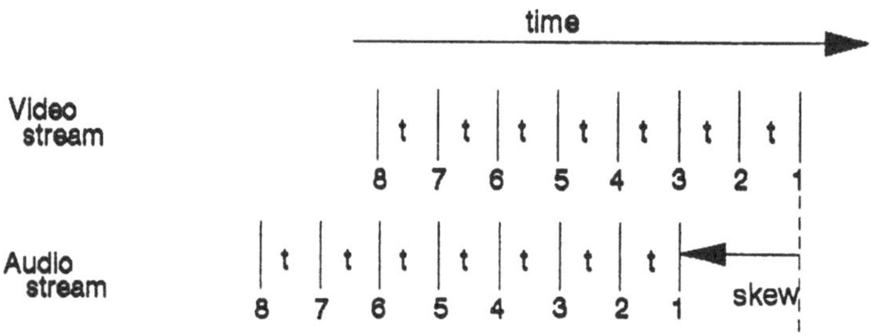

Figure 5 - Performance Metrics

CCITT in its work on broadband systems (7) defines <u>conversational</u> services as "direct interactive, real time communication based on the exchange of a mixture of voice, video, still image and data files between two partners".

The same work defines a number of important <u>distributive</u> application classes as follows:

<u>Collection Services</u>	Monitoring of distributed sensor stations gathering information of various kinds (many-to-one communication).
<u>Messaging Services</u>	Indirect one-to-one communication services using intermediate storage facilities.
<u>Retrieval Services</u>	Individual requests to multimedia databases or information centers.

Distribution Services — One to many (multicast or Broadcast) communication services of HIFI, video, or HDTV programs with or without user presentation control.

For all of these applications, whether of the distributive or conversational type, once the communication is underway with an acceptable latency, there are two additional significant performance requirements. First, there are requirements for synchronization among the various information types which can range from a "coarse" synchronization such as in sequencing the transmission of various objects (image, voice, next image, next voice, etc.) to a more precise "fine" synchronization such as synchronizing a voice to the speaker's lip motion. (The term "skew" refers to the difference in presentation times between two related objects.) Also, since multimedia includes real time information (video and audio), there are performance limits on "jitter", which represents instantaneous variation in latency or statistical distribution of the per packet communication delays. These metrics are important to the user's perception of smooth continuous motion and sound. Several references are available on the definition of multimedia performance metrics (1), (2), (3), (4). The metrics latency, jitter, and skew are summarized by the diagram in Figure 5.

III. SYSTEM PERFORMANCE REQUIREMENTS

System performance requirements for the two types of multimedia application (Conversational and Distributive) can now be described in terms of the performance metrics, latency, jitter, and skew. Here, "system" refers to the entire multimedia system. For conversational applications, this includes the live audio/video sources (camera and microphone), the real time compression equipment, the networking subsystem, the decompression equipment and the destination video display and speaker. In distributive applications, the system is similar except an audio/video server containing precompressed information replaces the live audio/video sources and the real time compression equipment. Thus "latency", for example, includes all delays in the system including sampling, compression, networking, decompression and presentation.

Figure 6 summarizes our current view of multimedia system performance objectives. Conversational audio latency requirements are derived from experience with telephone and echo cancellation systems (8). Here, 150 milliseconds is taken as a safe limit which would provide acceptable voice interaction even in the presence of impaired echo control circuits. Physiologically, the eye is more forgiving than the ear, however, for lack of better data, we assume a video latency requirement equal to audio requirements in conversational applications. In distributive applications, we can relax the latency requirements to a value of one second which is generally consistent with objectives for interactive computer systems.

Some of the skew objectives shown in Figure 6 are based on human factors measurements, others are the best judgments of various experts in the field. The most challenging skew requirement ("Fine Skew") is to synchronize voice with a speaker's lip motion. This requirement, sometimes called "lip sync" is important in applications involving close up views of a speaker ("talking head"). In real life, light

Media Combinations		Application Type	
		CONVERSATIONAL (real time, interactive,P2P)	DISTRIBUTIVE (e.g. Server to Client)
Audio Alone		Latency < 150 ms	Latency < 1000 ms
Video Alone		Latency < 150 ms	Latency < 1000 ms
Audio + discrete information: text,image, and/or graphics.		Audio requirements same as above PLUS: "Coarse Skew" - a coarse synchronization of audio with discrete information as in voice-annotated text or in a slide show with voice descriptions. Skew < 1000 ms	
Audio with Video	"Coarse Skew" Applications--	Audio and Video requirements same as above PLUS: coarse synchronization between audio and video as with narration, background music, or when speaker points to a visual object. Skew < 200 ms	
Audio with Video	"Fine Skew" Applications--	Audio and Video requirements same as above PLUS: fine synchronization as with "lip sync" or sound effects. audio advance of video < 20 milliseconds audio delay following video < 120 milliseconds	

Figure 6 - Multimedia System Performance Objectives

travels much faster than sound. Therefore, we are trained to accept a small delay between receipt of video information and the related sound. The opposite effect, perceiving sound before the related video image is much less tolerable. Research by Cooper (9) indicates that sound in advance of video by 20 milliseconds or more is disturbing to the viewer. Preliminary IBM measurements (10) have indicated that sound should not precede video by more than 40 milliseconds, nor lag video by more than 120 milliseconds. Within this range, some users may perceive a lack of lip sync, however, this will generally not be considered annoying, according to measurements conducted on a small sample of users.

Figure 7 summarizes latency and jitter objectives identified by an IEEE 802.11 wireless LAN committee on market requirements (11). It is interesting to note that the performance objectives vary with the audio and video quality level suggesting that a user dealing with high quality presentations may be less tolerant of distortions due to latency and jitter than users dealing with lower quality presentations. Also, it is important to recognize that Figure 7 represents only the Local Area Network subsystem, in contrast to the total system view in Figure 6.

Data Transmission Latency - milliseconds (1)

Application	MEAN	STND. DEV.	MAXIMUM*
64 Kbps Video Conference	30.0 ms.	100.0 ms.	250.0 ms.
1.5 Mbps MPEG NTSC Video	5.0	5.0	11.0
20 Mbps HDTV Video	0.8	0.8	0.8
16 Kbps Compressed Voice	30.0	100.0	250.0
256 Kbps MPEG Voice	7.0	7.0	16.0

(1) Reference: IEEE P802.11/92–20 Appendix I, Functional Requirements, dated January 1992.

* Note: "maximum" can be interpreted as 99.9th percentile, see text.

Figure 7 - LAN Objectives for Conversational Multimedia

The mean latency objectives shown in Figure 7 represent time from when a packet is submitted for transmission at the source MAC interface until the completion indication is given by the destination MAC interface between two stations within the same LAN segment.

The "jitter" characteristic of communications performance is represented in this work by two metrics, the standard deviation of packet latency, and the maximum tolerated packet latency. Packets delayed by more than this amount of time are considered by the application to be the same as lost packets. Hence, this metric is related to packet loss probability. The IEEE work (11) targets a maximum packet loss rate of 0.001 (one lost packet per thousand packets transmitted on the LAN). With this objective, and ignoring other losses, the maximum tolerated latency can be interpreted as the 99.9th percentile of packet latency.

IV. NETWORKING SUBSYSTEM OBJECTIVES

The two key attributes of the network in a multimedia system are bandwidth and latency. Sufficient bandwidth is required to support the aggregate Throughput Demand of the multiple sessions which share the network, without significant queueing and with minimal variation in communication delays. In cases of shared physical media, it can be shown that variance in communication delays increases with network utilization. Hence, for a given level of traffic, greater bandwidth leads to lower utilization and less "jitter" in communication. The latency of the communication network should be small compared to the allowable total system latency. With this, most of the allowable system latency can be budgeted to other functions such as compression and decompression, perhaps enabling a reduced Throughput Demand or a higher quality audio and video presentation. Networking design characteristics which can minimize latency through the network include prioritization of multimedia traffic over data traffic, managing bandwidth to limit total network utilization, and communication with fast packet switching techniques (fixed cell size and hardware based switching). Also, innovative congestion control methods may, for example, tradeoff video and audio resolution in real time to minimize communication latencies.

In addition to high bandwidth and small latency, the network's contribution to jitter and skew should be as small as is economically feasible. This, however, may be of secondary importance, because it may be less costly to compensate for jitter and skew with buffering and logic at the destination, than to design minimal jitter and skew into the network. Approaches to minimizing jitter use a "smoothing buffer" to accumulate a supply of audio and video packets which can be clocked at a fixed ("isochronous") rate to the destination's audio/video equipment. Of course, this smoothing takes time and adds to system latency, which puts further emphasis on the need for minimal latency in the network.

IV. SUMMARY

Multimedia applications are diverse, both in terms of throughput demand and in the performance requirements (latency, skew, jitter) placed on the networking system. The specification of networking performance objectives for a particular application, therefore, requires an analysis of several applications variables including the combinations of media, information update rates, audio and video quality requirements, compression capabilities, synchronization (skew) requirements and whether the application is conversational or distributive. Networking performance requirements for several representative applications are described in this report.

The primary role of the networking subsystem in multimedia applications is to provide high bandwidth and minimal latency. Networking techniques such as traffic prioritization, bandwidth management, and fast packet switching will contribute to success of the networking system in meeting the multimedia requirements.

REFERENCES

(1) T. D. C. Little and A. Ghafoor, "Network considerations for distributed multimedia object composition and communication," *IEEE Network Magazine*, Nov. 1990.

(2) D. Ferrari, "Client requirements for real time communication services," *IEEE Communications Magazine*, Nov. 1990.

(3) M. M. Mourad, "Some issues in the Implementation of Multimedia Communication Systems", IBM Research Report RC 15913 (#70743), Sept. 1990.

(4) D. B. Hehmann, M. G. Salmony, and H. J. Stuttgen., "Transport services for multimedia applications on broadband network," *Computer Communications*, Vol 13, no. 4, May 1990.

(5) J. D. Russell, "Trends in throughput demand," Fibertour/ Computernet 90 Conference, Boston, Ma., October 1990.

(6) J. T. Johnson, "Videoconferencing not just talking heads," *Data Communications Magazine*, Nov. 1991.

(7) CCITT Study Group XVIII, Broadband Task Group, Part C of the Report of the Seoul Meeting, 1/25/88, CCITT COM XVIII-R55(C)-E, Feb. 1988.

(8) CCITT Fascicle III.1-Recommendation G.114, Melbourne 1988.

(9) J. Carl Cooper, "Video-to-Audio Synchrony Monitoring and Correction", Society of Motion Picture and Television Engineers (SSMPTE) Journal, September 1988.

(10) IBM Hursley Human Factors Internal Study Note 38 R. Jackson et al, August 22, 1990.

(11) IEEE Computer Society, Standards Working Group IEEE P802.11, Document #92-20, January 28, 1992.

TRAFFIC MEASUREMENTS ON HIPPI LINKS IN A SUPERCOMPUTING ENVIRONMENT

Hung-Chang Kuo, Arne Nilsson

Center for Communication and Signal Processing
Department of Electrical and Computer Engineering
NCSU, Raleigh, N.C.

Dan Winkelstein

Center for Communication
MCNC, RTP, N.C.

Laura Bottomley

Department of Electrical Engineering
Duke University, Durham, N.C.

INTRODUCTION

In the future high speed networks, e.g., Broadband ISDN and Gigabit networks, we can expect three major types of traffic; namely, the variable bit rate data traffic, the variable bit rate video traffic, and the constant bit rate service traffic. The three types have different traffic characteristics and quality of service requirements. The variable bit rate data traffic is extremely bursty, and the connection between the source and the destination operates at its peak bandwidth, or is quiescent. The variable bit rate video traffic is not as bursty as the variable bit rate data traffic, but is more delay-sensitive, as the temporal relationship of the information is significant. The constant bit rate service traffic has a constant cell stream, and is sensitive to delay and delay variation, as the data need to be delivered timely. Its sensitivity to cell loss will depend on the application of the service.

To analyze the network performance, first, we need to understand the traffic characteristics, and model the traffic source realistically; and second, based on this traffic source model analyze the network performance. Without good traffic measurement, none of these can be done.

The modeling of the constant bit rate service traffic is trivial. The variable bit rate video service is a new service which is expected to be of great importance in the future high speed networks. The study by Nomura, Fujii, and Ohta [NO89] is a very good reference. Our study will focus on characterization and analysis of the variable bit rate data traffic on very high speed links in a super-computing network environment.

There are several traffic measurements and analyses on local area networks and wide area networks published recently [JA86], [GU91], [FO91], [LA90], [BO91]. Although the measurement data are coming from networks of higher transmitting speed than traditional packet switching network, the highest transmission rate among these measured networks is only 20 Mbps. These measurements reveal a lot of possible high

Asynchronous Transfer Mode Networks, Edited by Y. Viniotis
and R.O. Onvural, Plenum Press, New York, 1993

speed network characteristics, e.g., high burstiness, high data correlation and so on. There is, however, no substitute for the measurement on the real high speed networks, some key characteristics of high speed network traffic may be totally new or quite different from the existed networks.

MEASUREMENT ENVIRONMENT

The traffic measurements presented here are taken from HiPPI links in a super-computing environments at the Microelectronics Center of North Carolina (MCNC). The data transmitting rate of HiPPI links is 800 Mbps. The traffic collection tool is called HIPPI Link Data Analysis system (HILDA) [WI91], and was developed by MCNC for real time traffic analysis of very high speed data network. The traffic measurement is from very high speed links in a super-computing environment, and it will help us fill up some gaps in very high speed traffic measurement and gain more insight of the characteristics of the high speed network traffic.

The HiPPI Interface

HiPPI stands for High Performance Parallel Interface , which is an ANSI X3T9.3 standard for very high speed data transmission [BH90]. It has two configurations. The first one uses a 32 bits bus and provides 800 Mbps data transfer. The second one uses a 64 bits bus and provides 1600 Mbps data transfer. In our measurement environment, we use the 800 Mbps configuration. HiPPI link operates in the simplex mode. When a full duplex transmission is needed, two HiPPI links are required. HiPPI interface is suitable for point-to-point physical connection or may be used in a circuit switch environment. It consists of four parts, which correspond to the OSI model's layer one and layer two. The following is a brief description of the HiPPI standard.

1. HiPPI-PH specifies mechanical, electrical, and signalling protocol.
2. HiPPI-FP specifies the framing protocol. It can make large block data transfers with framing to split the data into smaller burst, support identifiers for multiple upper layer protocols, provide error notification for damaged data, etc.
3. HiPPI/IPI specifies intelligent peripheral interface with the higher layer protocol.
4. HiPPI-MP specifies memory port.

Figure 1 shows the relationship between the components and the layers.

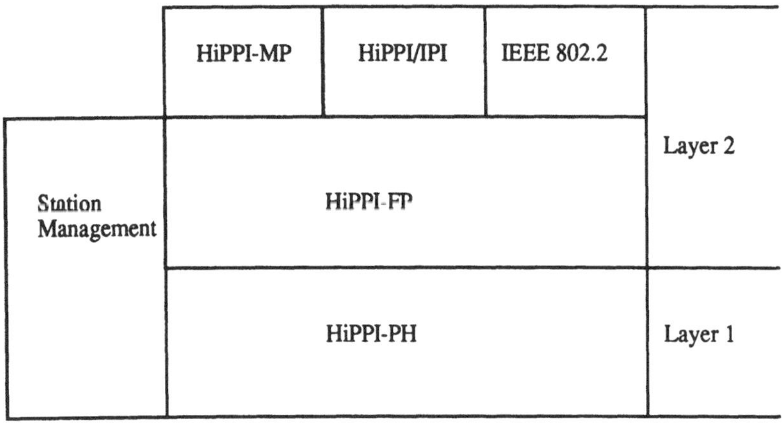

HiPPI-MP
HiPPI/IPI
IEEE 802.2
Layer 2
Station Management
HiPPI-FP
HiPPI-PH
Layer 1

Figure 1 HiPPI standards composition.

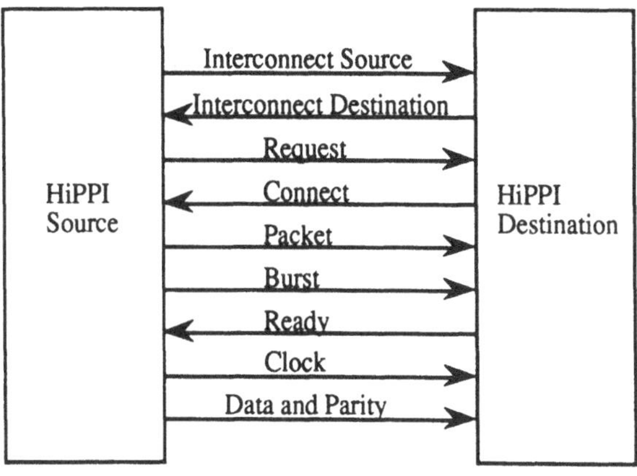

Figure 2. HiPPI signals.

HiPPI-PH Signalling Mechanism. Because HiPPI signalling mechanism in HiPPI-PH has some impacts on traffic measurement results, we'll give further discussion of it. Figure 2 illustrates the signalling procedures between a source and a destination.

1. Interconnect Source and Interconnect destination signals are used to check hardware readiness.
2. Request and Connect signals are used to set up connection.
3. Packet signal is used to delineate the beginning and the end of a packet. There is no limit on the size of a packet.
4. Each packet is comprised of bursts, e.g., 1K bytes of data; thus, we may have several bursts within a packet transmission.
5. Ready signal is used for flow control. The destination will assert the ready signal once for each burst it is prepared to receive. If, on the other hand, if the destination is busy, it can withhold or slow down sending ready signal.
6. Clock signal is used for clocking and the clock rate is 25 MHz.
7. Data and Parity signal is used for error detection. Two mechanisms have been used to detect errors in transmission. The first one uses parity bit associated with each byte, and the second one uses Length/Longitudinal Redundancy Code (LLRC). Together, they guarantee to detect all three-bit (or fewer) errors.

HiPPI Link Data Analysis System (HILDA)

The hardware of the HILDA system is a single board on the VME bus of a SUN4 host. Figure 3 shows the location of HILDA in our measurement environment. The board in conjunction with the kernel and application software (see Figure 4) comprise the HILDA system, which operates in the following functional modes.

1. Traffic analysis mode

   In this mode HILDA is completely transparent to the HiPPI host, and the packet delay across the board is negligible. Real-time statistics on packet length, packet interarrival time, error rate, ready lookahead, etc. can be collected. The user may display the data in graphical form in real time or just collect raw data for further processing. In our traffic measurements, we collect the raw data of packet interarrival times and packet length.

   The time resolution of traffic collected can be as small as 40 ns. This fine resolution is necessary for traffic collection in a super-computing environment. In a 800 Mbps network, if the packet size is not less than 32 bits, we can't have two packets in the same time unit. Indeed, the minimum packet size in our measurement is 4 words (32 bits), and the measurement doesn't miss any packets going through HiPPI links. But, rarely, because of the high time resolution, we did have interarrival time overflow occurrence. That is, when interarrival time is greater than x'fffff0'*40ns (about 10737 ms), we'll have interarrival time overflow occurred. However, HILDA does have other time resolutions, up to 10.24μs.

2. HiPPI host mode

   While in this mode, HILDA provides a full duplex HiPPI connection over two simplex links for the VME host. This mode allows the VME host to carry out link stress test. Link stress test is necessary to test the robustness of the link protocol and the network design. Since our main focus is to understand the normal traffic pattern, also, due to the storage limitation, it is almost impossible to save such big volume of data in link stress test.

3. Programmable error insertion mode

   It is expected that the error rate in a high speed network is very low. This HILDA mode provides a controllable error insertion in test environment, so that the impacts of error condition on traffic measurements can be evaluated. The inserted error rate can be as high as 1 bit error in 10,000 bits.

4. Data Extraction mode

   In this mode, we can collect samples of packet for up to 16 Kbytes long. Since HiPPI is a point-to-point protocol, we know the source application and destination application right away. The contents of the packet can serve as a data integrity check, but doesn't offer us any more information, so usually we don't run this mode.

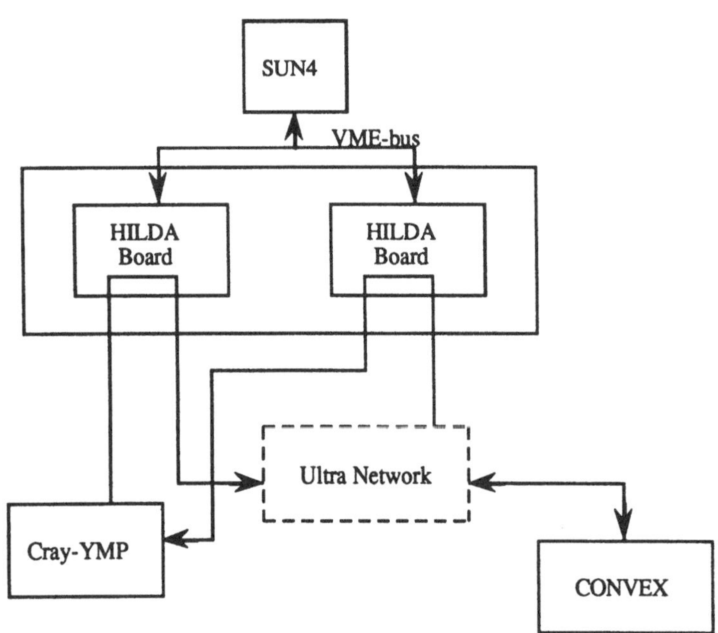

Figure 3. Measurement environment.

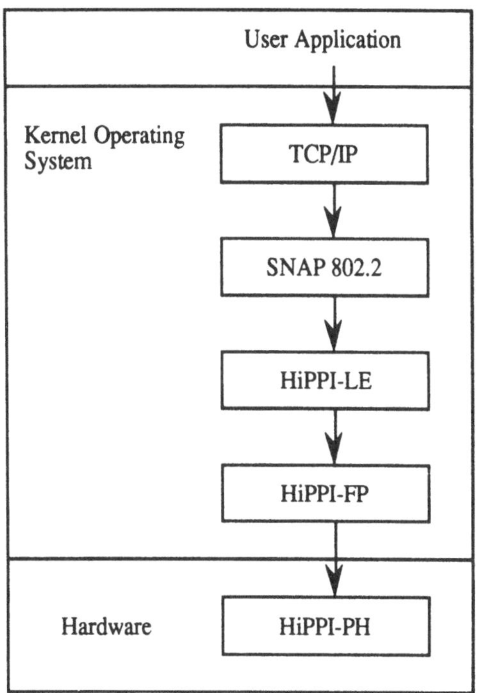

Figure 4. Protocol stack used by HiPPI interface.

Application Programs Used in Measurement Environment

Figure 4 shows the protocol stack used by the HiPPI interface. The kernal operating system is modified to support both network layer and transport layer. On top of the protocol stack is TCP/IP, with SNAP 802.2, HiPPI-LE (HiPPI Link Encapsulation Layer [HL90]), HiPPI-FP, and HiPPI-PH under it. When SUN4 is the host, the whole protocol stack will be used. When CRAY-YMP is the host, the protocol stack is simpler; SNAP 802.2 and HiPPI-LE are not used. When the application is UFTP (see below), it is even simpler, TCP/IP is not used.

There are three application programs used in HiPPI links measurements environment. The first application program is TSOCK, which is an internal network performance test tool used within the MCNC communication department. When two TSOCKs try to communicate with each other, first sockets are being set up, then the communication can start. The second application program used is FTP (File Transfer Protocol), which is a standard for file transfer in TCP/IP environment. FTP uses the idea of socket too. The third application program is UFTP (File Transfer Protocol using ULTRA interface i.e., HiPPI interface), which is a modification of FTP and is used in the test environment within MCNC. UFTP is used in file transfer too. But it can use HiPPI interface directly and no need to use TCP/IP protocols.

INTERARRIVAL TIMES ANALYSIS TECHNIQUES

In the analysis of interarrival times, we may view the collected data of interarrival times as a series of events. There are two approaches to analyze a series of events. The first approach treats the series of events as a non-stationary process, i.e., there is a temporal trend of the events, or, in other words, there is a trend for the event occurrence rate. One example of the event occurrence rate is the arrival rate l in the Poisson process. If a trend of the event occurrence rate can be found, the occurrence of the future event may be predictable. The second approach assumes that the asymptotic values of the event occurrence rate, first moment, second moment, third moment of the events, etc. do exist, when time reaches infinity. That is, there is a stationary state for those parameters. Cox D.R. [CO66] calls this kind of processes stationary processes.

The second approach can be seen in a number of Markovian-related process analyses, e.g., the analyses using Poisson process, IBP (Interrupted Bernouli Process), IPP (Interrupted Poisson Process), MMPP (Markov Modulated Poisson Process), etc. This approach has been used broadly in modeling data traffic in lower speed networks satisfactorily. However, the traffic on a very high speed network, like that on a HiPPI link, is extremely bursty and highly correlated, the stationarity assumption may not hold.

Before we can decide which approach to take, we really need to know the stationarity property of the process. One way to find it is doing the trend analysis of the event occurrence rate (trend analysis, in short).

Trend Analysis of the Event Occurrence Rate

The trend analysis algorithm used here is based on Cox D. R. [CO66]. It uses the least-squares regression method to analyze trend of events. This method is flexible and works reasonably well under fairly weak assumptions about the structure of the series of events.

There are some theoretical problems related with the algorithm we are going to use. We'll discuss them first.

Log Chi-squared Distribution. Let T_{n0} be the sum of n0 identical and independent exponentially distributed random variables. The occurrence rate of the exponential distribution is assumed to be λ. We know that T_{n0} has a Gamma distribution with p.d.f.

$$\frac{\lambda(\lambda t)^{n0-1}e^{-\lambda t}}{(n0-1)!} \quad (t \geq 0).$$

For large n0, T_{n0}, by the Central limit theorem, is asymptotically normally distributed. Thus $(\lambda T_{n0} - n0)/\sqrt{n0}$ has a standardized normal distribution. Although the random variable T_{n0} has this nice feature, it will not be used in our trend analysis. We will use the random variable $\ln(T_{n0})$ instead, and the reason for this will be discussed next.

We know that $2\lambda T_{n0}$ has the chi-squared distribution with 2n0 degree of freedom; symbolically

$$2\lambda T_{n0} \in X_{2n0}{}^2.$$

Thus,

$$\ln(T_{n0}) \in -\ln(2\lambda) + \ln(X_{2n0}{}^2),$$

now we have the log chi-squared distribution , and the analytical results derived by Bartlett and Kendall [BA46] can be used. The results are

$$E(\ln T_{n0}) = \psi(n0) - \ln\lambda, \qquad \mathrm{Var}(\ln T_{n0}) = \psi'(n0), \tag{3.1}$$

where

$$\psi(x) = \frac{d\ln\Gamma(x)}{dx}$$

is the digamma function [DA33],

$$\Gamma(x) = \int_0^\infty t^{x-1} e^{-t}\, dt$$

is the Gamma function, and $E(\ln T_{n0})$ and $Var(\ln T_{n0})$ are the mean and variance of $\ln T_{n0}$ respectively.

Bartlett and Kendall obtained the approximation of $E(\ln T_{n0})$ and $Var(\ln T_{n0})$ by empirically modifying the asymptotic expansion of $\psi(n0)$ and $\psi'(n0)$ into the following result:

$$E(\ln T_{n0}) \simeq \ln\left(\frac{n0}{\lambda}\right) - \frac{1}{2n0 - \frac{1}{3} + \frac{1}{16n0}},$$

$$Var(\ln T_{n0}) \simeq \frac{1}{n0 - 0.5 + \frac{1}{10n0}}. \quad (3.2)$$

The approximation is fairly accurate; the error is only 1 at n0=1, and when $n0 \neq 1$ the $1/n0$ term can be omitted.

Regression Analysis of Intervals. Based on the results of the previous section, we can start the discussion about the regression analysis. First, we make the approximation that the event occurrence rate λ is effectively a constant λ_i within a time period y_i. Here y_i is the i-th aggregate interval, which is formed by grouping k successive intervals together. From (3.2), we have

$$E(\ln(y_i)) = -\ln(\lambda_i) + e_k,$$

$$Var(\ln(y_i)) = v_k, \quad (3.3)$$

where e_k and v_k are known constants and are independent of λ_i, with

$$e_k \simeq \ln(k) - \frac{1}{2k - \frac{1}{3} + \frac{1}{16k}},$$

$$v_k \simeq \frac{1}{k - 0.5 + \frac{1}{10k}},$$

respectively. Second, we define the independent variable z_i as the time at the midpoint of the aggregate interval y_i. Assume the event occurrence rate λ_i, which is the dependent variable, is a function of z_i, and can be expressed as

$$\lambda_i = e^{-(\alpha + \beta_1 z_i + \beta_2 z_i^2 + + \beta_3 z_i^3 \ldots)}$$

The assumption of λ_i in this form can cover both trend existed case and no trend existed case. When $\beta_1, \beta_2, \beta_3 \ldots$ are all zeros, $\lambda_i = e^{-\alpha}$ is a constant and no trend exists; otherwise we have a trend. Taking logarithm of λ_i, we have

$$\ln(\lambda_i) = -(\alpha + \beta_1 z_i + \beta_2 z_i^2 + \beta_3 z_i^3 \ldots). \quad (3.4)$$

Substituting (3.4) into (3.3), we have

$$E(\ln(y_i)) = \alpha' + \beta_1 z_i + \beta_2 z_i^2 + \beta_3 z_i^3 \ldots \quad (3.5)$$

where $\alpha' = \alpha + e_k$.

From (3.3) and (3.5), we can find the merits of using the log chi-squared distribution immediately. First, the variance of $\ln(y_i)$ is a constant, this is an assumption needed in the regression analysis. Second, the mean of $\ln(y_i)$ is a linear function, thus the least-squares analysis can be applied easily.

We know, theoretically, that y_i, by the Central limit theorem, will approach a normal distribution when the grouping factor k becomes large. In the regression analysis, it is assumed that the mean óf the dependent variable is normally distributed. So, we may ask if $E(\ln(y_i))$ is close to normal distribution when the grouping factor k becomes large. The answer is no. Instead, the log chi-squared distribution has a moderate negative skewness and tends rather slowly to normality as $n0 \to \infty$. Fortunately, this non-normality effect only produces a relatively unimportant distortion of significance levels and confidence coefficients in the regression analysis.

The method stated above can be used in the trend analysis of a process in which the aggregate intervals are mutually independent and with constant coefficient of variation. In a more general process with trend, however, there will be correlation between different $\ln(y_i)$s. Nevertheless, as long as these correlations are not too large, we shall get reasonable answers from the regression analysis. That is, the regression methods are by no means tied to an underlying exponential distribution of intervals.

Since we like the correlation between aggregate intervals to be as small as possible, we have to choose the grouping factor k carefully. Cox has suggested that the grouping factor k should be at least 4. Here, we use an autocorrelation coefficients series of different lags to decide k, i.e. when the autocorrelation coefficients series is approaching something stable and negligible, e.g., 5%, we pick the k-value as the grouping factor. This way, we will have those more correlated intervals in the same group and those not so much correlated in the next new group. If the grouping factor can't be determined this way, e.g., when the autocorrelation coefficients series does not ever approach zero at all, the experience from other measurement data will be used.

The next step will be using the least-squares analysis to find those parameters, α, β_1, ..., and it will be discussed in the next section.

Least-Square Method and Orthogonal Polynomials. When using least square method to find those parameters, α', β_1, ..., we may have two approaches: one is using solving simultaneous linear equations technique, and the other is using orthogonal polynomials technique. We'll have discussions of both as follows.

Finding Parameters by Solving Simultaneous Linear Equations. Let y be the mean of the logarithm of the aggregate interval time, i.e. $E(\ln(y_i))$, and

$$y = k_0 + k_1 x + k_2 x^2 + \ldots + k_m x^m \quad (3.6)$$

Here, k_0, k_1, k_2, ..., k_m correspond to α', β_1, ..., β_m, and x corresponds to z_i as in (3.5). Clearly y is a funcion of x, or, in other words, a function of z_i; symbolically, $y = y(z_i) = y(x)$.

Define S as the sum of the square of $(y_i - y(z_i))$, i.e.,

$$S = \sum_{i=0}^{n} \left(y_i - \sum_{j=0}^{m} k_j x^j \right)^2 ,$$

where n plus one is the total number of aggregate intervals.

To get the least square sum of S, we take partial derivatives of S with respect to k_j, j=0, 1, 2, ...,m , and make them zero; symbolically

$$\frac{\partial S}{\partial k_j} = 0, \qquad j = 0, 1, 2, ..., m.$$

By using the matrix notation, we have

$$\begin{bmatrix} a_{00} & a_{01} & \cdots & a_{0m} \\ a_{10} & a_{11} & \cdots & a_{1m} \\ . & . & \cdots & . \\ . & . & \cdots & . \\ a_{m0} & a_{m1} & \cdots & a_{mm} \end{bmatrix} \begin{bmatrix} k_0 \\ k_1 \\ . \\ . \\ k_m \end{bmatrix} = \begin{bmatrix} b_0 \\ b_1 \\ . \\ . \\ b_m \end{bmatrix},$$

where $a_{jk} = \sum_{i=0}^{n} x^{j+k}$, and $b_k = \sum_{i=0}^{n} x_i^k y_i$, j=0, 1 ... m, k=0, 1 ... m .

This method works fine, when the degree of the polynomial, m, is not too large. When m becomes larger, ill conditions of the m simultaneous equations may occur; that is, we may not have a unique solution. Also, for every different value of m, we need to solve a set of m simultaneous equations, if the total number of aggregate intervals is large, which is the case in our interarrival time analysis, it is really time-consuming to calculate those elements of the matrixes i.e., a_{jk} and b_k.

Finding Parameters by Using Orthogonal Polynomials. This method, although more complicated, doesn't have the drawbacks mentioned above. First let's define orthogonal functions, $\Phi_m(x)$s m = 0, 1, 2, ..., as a set of functions satisfying the following condition:

$$\int_a^b \Phi_m(x)\, \Phi_n(x)\, dx = 0 \quad m \neq n .$$

If those orthogonal functions can be expressed as polynomials, we call them orthogonal polynomials. In our analysis, the orthogonal polynomials being used are Chebyshev polynomials.

Chebyshev Polynomials. Chebyshev polynomials are defined as follows:

$$T_r(x) = \cos(r \cos^{-1} x) \quad -1 \leq x \leq 1 \qquad r = 0, 1, 2, 3, ...,$$

that is,

$$T_0(x) = 1 ,$$

$$T_1(x) = x ,$$

$$T_2(x) = 2x^2 - 1 ,$$

$$T_3(x) = 4x^3 - 3x , \quad \text{and so on.}$$

It has a nice recursive relationship.

$$T_{r+1}(x) - 2xT_r(x) + T_{r-1}(x) = 0 \quad r=1,2,3, \ldots,$$

If we choose

$$\overline{x}_i = \cos\frac{(2i-1)\pi}{2(m+1)} \qquad i = 1, 2, \ldots, m+1, \tag{3.7}$$

then we have the following orthogonal polynomial property

$$\sum_{i=0}^{m+1} T_k(\overline{x_i})T_j(\overline{x_i}) = \begin{cases} 0 & k \neq j \\ \frac{m+1}{2} & k=j\neq 0 \\ m+1 & k=j=0 \end{cases} . \tag{3.8}$$

It is clear that the independent variable we use, z_i, is not in the range of [-1, 1]. So, we make the following transformation of z_i into a new independent variable x_i, such that $-1 \le x_i \le 1$. The transformation is

$$x_i = \frac{z_i - \frac{(z_n + z_0)}{2}}{\frac{(z_n - z_0)}{2}} = \frac{2z_i - (z_n + z_0)}{(z_n - z_0)},$$

where z_0 and z_n correspond to the independent variables on the first interval and the last interval respectively. From this transformation, we can see $x_0 = -1$, $x_n = 1$, and all other x_is have values in (-1,1).

Lagrange Interpolation. For a given Chebyshev polynomial of degree m, first, we have to find the $\overline{x_i}$s from (3.7), so that the orthogonal polynomials properties of (3.8) can be used. Then, because those $\overline{x_i}$s may not have measured $y(\overline{x_i})$s corresponding to them, we have to estimate them.

If $y(\overline{x_i})$ already exists, we don't have problem; if not, we can use interpolation technique to find it. Here we use Lagrange interpolation technique [KU71], because it uses multiple neighboring points to find the interpolated $y(\overline{x_i})$, so better estimate of the interpolated value can be obtained.

The Lagrange interpolation formula is summarized as below:

$$y(\overline{x_i}) = \sum_{k=0}^{j} y_k \prod_{i=0}^{j} \frac{\overline{x} - x_i}{x_k - x_i} \qquad \text{and } i \neq k, \tag{3.9}$$

where (x_k,y_k), k = 0, 1, 2, ... j, are j plus one pairs of known data used in finding the interpolated value $y(\overline{x})$. In our implementation, we use j = 20, and the results are fairly accurate.

Least Square Method Using Chebyshev Polynomials. From (3.9), we can obtain all $y_i(\overline{x_i})$s. Let y be the mean of the logarithm of the aggregate interval times, i.e. $E(\ln(y_i))$, and is represented as combinations of Chebyshev polynomials of different degrees; that is,

$$y = C_0T_0(\overline{x_i}) + C_1T_1(\overline{x_i}) + C_2T_2(\overline{x_i}) + \ldots + C_mT_m(\overline{x_i}). \tag{3.10}$$

Define S as the sum of the square of $(y_i - y)$, i.e.,

$$S = \sum_{i=0}^{n} \left(y_i - \sum_{j=0}^{m} C_j T_j(\overline{x_i}) \right)^2 ,$$

where n plus one is the total number of aggregate intervals.

Take partial derivatives of S with respect to C_j, j=0, 1, 2, ..., m , and make them zero; symbolically

$$\frac{\partial S}{\partial C_j} = 0, \qquad j = 0, 1, 2, ..., m.$$

By using the matrix notation, we have

$$\begin{bmatrix} t_{00} & t_{01} & \cdots & t_{0m} \\ t_{10} & t_{11} & \cdots & t_{1m} \\ . & . & \cdots & . \\ . & . & \cdots & . \\ t_{m0} & t_{m1} & \cdots & t_{mm} \end{bmatrix} \begin{bmatrix} c_0 \\ c_1 \\ . \\ . \\ c_m \end{bmatrix} = \begin{bmatrix} d_0 \\ d_1 \\ . \\ . \\ d_m \end{bmatrix},$$

where $t_{jk} = \sum_{i=0}^{m} T_j(\overline{x_i}) T_k(\overline{x_i})$, and $d_k = \sum_{i=0}^{m} y_i(\overline{x_i}) T_k(\overline{x_i})$, j = 0, 1, ..., m, k = 0, 1, ..., m.

We can see, from the properties of orthogonal polynomials in (3.8), that all $\sum_{i=0}^{m} T_j(\overline{x_i}) T_k(\overline{x_i})$ terms are zeros when $k \neq j$. This quickly leads to

$$C_j = \frac{\sum_{i=0}^{m} y_i(\overline{x_i}) T_j(\overline{x_i})}{\sum_{i=0}^{m} T_j^2(\overline{x_i})} \qquad j = 0, 1, 2, ..., m. \tag{3.11}$$

In our interarrival time analysis, although may sacrifice some accuracy, we try to keep our traffic model as simple and representative as possible. So, we only fit polynomial up to m=10. The computation time is not an issue in this approach, even m becomes large, it doesn't take long to compute C_j in (3.11).

Index of Dispersion for Intervals

The index of dispersion for interarrival intervals can give us an idea of whether a stationary process is renewal or not. Let us assume that the traffic process is second order stationary.

Let $X_1, X_2, ..., X_k$ be a sequence of interarrival times. Define variance function V_k as

$$V_k = \mathrm{var}(X_1 + X_2 + ... + X_k)$$

$$= k\,\mathrm{var}(X) + 2\sum_{j=0}^{k-1} \sum_{h=1}^{j} \mathrm{cov}(X_j, X_{j+h})$$

$$= kV_1 + 2\sum_{j=0}^{k-1} (k - j)C_j \; . \quad (3.12)$$

In (3.12) we have used the second order stationarity assumption; that is,

$\mathrm{var}(X_1) = \mathrm{var}(X_2) = ... = \mathrm{var}(X_k) = \mathrm{var}(X) = V_1$, and

$\mathrm{cov}(X_h, X_{j+h}) = \mathrm{cov}(X_{j+h}, X_h) = C_j$,

where C_j is the auto-covariance function of lag j.

The index of dispersion for intervals, J_k, is defined as

$$J_k = \frac{V_k}{kE(X)^2} \; . \quad (3.13)$$

For a renewal process, we have a series of independent and identical random variables for intervals; so all C_js are equal to zero. That is, $V_k = kV_1$. This leads to J_k equal to $C(X)^2$ for all k, where C(X) is the coefficient of variation; that is, for a renewal process, J_k equals to $C(X)^2$ for all k. Poisson process is a special case of renewal process, and it can be shown easily that J_k equals to 1.

Thus, after the trend analysis, if we find it is a stationary process, we can use the index of dispersion for intervals, J_k, to determine if it is a renewal process. If it is not a renewal process, we may want to try those non-renewal but stationary processes, e.g., MMPP.

MEASUREMENT RESULTS AND TRAFFIC CHARACTERISTICS ON HIPPI LINKS

Prior to the widespread usages of local-area networks (LAN's) and, more recently, the planning of metropolitan-area networks (MAN's), bandwidth was a relatively expensive and often scarce resource in a communication system. The performance bottleneck was the network. The throughput available to application programs was usually limited by the network. With the advent of LAN's and MAN's, the performance bottleneck has shifted from the network to the processing required to execute communication protocols in workstations and servers. Only a very small fractions of the bandwidth of a communication network is used. Traffic is extremely bursty and sometimes highly correlated. This kind of effect is even more vivid on very high speed networks.

The traffic measurements on HiPPI links have clearly shown this kind of characteristics. We have very low link utilization for data transmission, and immediately after bursts of data transmission follows a a very long idle time, which relates to the communication protocol and application program processing times. The coefficient of variation, which can indicate the traffic burstiness, is very high. The autocorrelation coefficients of different lags, which indicates the correlation among data, is significant.

Because of the complete reversal of networking environment, we can expect that the traffic pattern on a very high speed network will be different from any traffic pattern encountered in our current existed networking environment.

Preliminary Discussion of Traffic Characteristics and Statistics on HiPPI Links

In our measurement, interarrival time is defined as the interval from the start of a data transmission to the start of the next data transmission. Thus for each interarrival time, there is one packet transmission associated with it.

From the measurement results, we find two kinds of packets: those with very short length, in ten's or less of words, and those with relative long length, in thousand's of words. The long ones correspond to the data packets and the short ones correspond to the control packets, i.e., these packets are generated by the communication protocols or servers. We can see that the processing times of the communication protocols and servers occupy a significant portion of total data transmission time. We also find that the time to establish or disconnect a connection between a source application and a destination application is also long. Another significant factor that affects the total transmission time is the operating system effect. In a multi-programming environment, many processes are competing for the host resources. Since the data transmission is just one of the processes, it is subjected to the rule of the operating system. The operating system preemption time used to be negligible when the communication network is the bottleneck. This is no longer true in the high speed network. In fact, in our super-computing environment, the system preemption time from the supercomputer is the longest when comparing with data transmission time, protocol processing time, application program processing time, etc.

Definitions of Some Statistics. The first step to get some ideas about the traffic characteristics is calculating the traffic statistics. We have calculated the following traffic statistics.

1. Mean and coefficient of variation for interarrival times and packet length,
2. Histograms for interarrival times and packet length. These histograms can help us see the distribution of interarrival times and packet length.
3. Autocorrelation coefficients of different lags for interarrival times. They are essential in the analysis of correlations among data.
4. HiPPI link utilization.

The following are some of the definitions of the statistics mentioned above.

Coefficient of variation is defined as

$$C = \frac{\sqrt{Var(.)}}{E(.)} ,$$

where E(.) and Var(.) are the mean and variance for interarrival times or packet length respectively.

Autocorrelation coefficients of lag k is defined as

$$C_k = \frac{E[(T_i - E(T))(T_{i+k} - E(T))]}{E(T)^2},$$

where T_i and T_{i+k} are random variables for interval i and i+k respectively.

Finally, we define the HiPPI link utilization

$$U = \frac{\text{(Total bits transmitted)}}{\text{(800 Mbps)(Duration of the transmission)}} .$$

The Outlying Interarrival Times Effects on Traffic Statistics. From the interarrival times histograms, we can see that there is a small percentage of extremely large interarrival times in every traffic measurement. Intuitively, these outliers seem to have big impacts on mean, coefficient of variation, and autocorrelation coefficients for interarrival times. If we want to look into these outliers effect more closely, we need to define a cut-off point for outlying interarrival times, i.e., any interarrival times greater than this cut-off value will be treated as an outlying interarrival time. After comparing all interarrival times histograms, we pick 42 ms as the outlying interarrival times cut-off point. It is really an inspective approach. There are two reasons to pick it. First, 42 ms is about 64K 640ns-time-unit or 1024K 40ns-time-unit; both 4k and 64k are very nature choices in computer science field, and they are very efficient for computer computations too. Second, it is a conservative number for outlying cut-off point in our measurements. When this value is used, the outlying areas are ranging from less than 0.1% to about 2.5% in our measurements.

The Outlying Interarrival Times Effects on Mean and the Coefficient of Variation. In order to understand these outliers effect on mean and variance, we calculate the percentages that they contribute to the sum of total interarrival times and to the sum of total interarrival times squares; that is, the percentages of their contribution to the first and second moments of interarrival times. Table 1 shows the percentages of outliers, and their contributions to the first moment and the second moment. The outliers effects on the first moment are ranging from 22% to 72%, and on the second moment are ranging from 84% to 99.9%.

Table 1. Interarrival time outliers' percentages and effects on 1st moment and 2nd moment

Measurement number	1	2	3	4	5	6
Outliers percentage (%)	2.5	2.5	0.35	2.5	0.82	0.1
Outlier effect on 1st moment	0.30	0.32	0.23	0.72	0.38	0.40
Outlier effect on 2nd moment	0.84	0.95	0.97	0.99	0.99	0.999

Outliers' Locality Distribution and Possible Causes of Creating Outliers. The localities of the outliers not only have great effects on the value of autocorrelation coefficient, but also reveal the causes that the outliers are created. We, first, define the locality of an outlier as its corresponding interval number, e.g., if an outlier corresponds to the 100th interarrival times interval, its locality is 100. Then, from this definition, we can calculate the locality distance between two consecutive outliers and the mean locality distance.

In general, the outliers in our measurements have shown the following five kinds of locality distributions:

1. Outliers cluster at the beginning of the measurement;

   They are possibly corresponding to the overhead to set up a connection between two application programs, for example, HiPPI interface processing, setting up socket, establishing connections between two application programs, etc. The reasonings are as follows. First, all outliers have control packets related with them, so we know they are not related with data transmission. Second, their localities are at the beginning. Third, we know, in the high speed network, that the connection setup time typically is much longer than data transmission time, and this is reflected in the big outliers at the beginning of the measurement period.

2. Outliers cluster at the end of the measurement;

   They correspond to the overhead to end a connection between two application programs. The arguments are almost the same as the previous type, except these outliers appear at the end of the measurement.

3. Extreme large outliers cluster in the middle of the measurement;

   This is probably due to the operating system effect from Cray-YMP. The reasonings are as follows. First, all outliers have control packets associated with them, so we know there are not related with data transmission. Second, the Cray-YMP's preemption mechanism, which swaps all current-processed processes out and swaps in all new processes, can contribute to a long idle time for the data transmission, if the data transmission is not finished before the preemption. Third, this kind of clustering only occurs when Cray-YMP is used.

4. Outliers cluster at the beginning of the measurement with the Cray-YMP's preemption involved.

   This one is similar to the previous one, but there are some differences. First, the transmission process, instead of being preempted, is waiting for the Cray-YMP's preemption, so that it can get served. Second, when this transmission finally gets served by Cray-YMP, we are going to have the first type of outliers clustering followed.

5. Very small group of outliers, e.g., 3 or 4, appears, now and then, in the middle of the measurement.

   There are several possible causes for them.

   a. Preemption from the SUN4 host.

      Because the preemption from the SUN4 host is not as drastical as the preemption from Cray-YMP, we can't distinguish them with other outliers. But we know they exist.

   b. The destination host is temporary timing out due to the overflow condition or having too many processes to process.

      In a super-computing environment, the destination host simply can't keep up with the pace of super-computer, so once in a while, the flow control mechanism will be activated to slow down the traffic.

   c. TCP/IP error recovery scheme gets started because of data error found during the transmission.

      In the high speed network, the error recovery time, which is much longer than data transmission time, can create outliers.

   All outliers have control packets associated with them, so we know they are not related to data transmission.

The causes for outliers, which we have discussed above, should occur rarely in a super-computing envoronment with 800 Mbps HiPPI links. Because data get transmitted so quickly, they don't stay in the system very long, thus the chances to be preempted are not high. Indeed, this is one of the criterion behind our choice of the outlying cut-off point.

To summarize the discussion above, we may say that these outliers are closely related with all of the background environment effects during the data transmission. Clearly these outliers have complete different properties from the time intervals used by data transmission. To analyze these outliers properties, we need to do the performance analysis of the operating system, HiPPI protocol, protocols used by servers, etc. Here, our focus is on the characterization of high speed network data traffic, thus we will have no further discussion about them.

One thing needs to be noticed is that there are always control packets in the middle of the data transmission, and they should be part of the data measurement. What we want to filter out is outliers which have completely different properties and needs different kind of analysis approach.

Measurement Results Analyses

In this section, we'll have several measurements study. For each measurement we have the following two analyses: 1) analysis using total measurement data i.e. total data case, and 2) analysis using only the measurement data without outliers i.e. without outliers data case, in short).

The physical setup of the test environment is given in Figure 3. The Ultra network serves as a gateway between two heterogeneous networks, i.e., one using HiPPI interface and the other using a different protocol. The VME bus, which is running at 20 these include mean, the coefficient of variation, autocorrelation coefficients of different lags, interarrival times histogram, packet length histogram, and HiPPI link utilization. In the second part, based on some of the results from the first part, we have interarrival times trend analyses, discussion of the index of dispersion for interarrival time, and some comparisons among different measurements.

Measurement Statistics. Tables 4-7 show the means, coefficients of variations for interarrival times, the means of packet length, and HiPPI link utilizations in all measurements for both total data and without-outliers cases. We can see that the decreases in both means and coefficients of variation for without-outliers cases, when comparing with total data cases, are significant. The link utilizations are low, typical for bursty data transmission. When the outliers are excluded, as expected, the link utilizations become higher.

Figures 5- 7 show the histograms of interarrival times. In all measurements, we can see a very long tail of outliers.

In Figures 8-10 the histograms of packet length are presented. The multi-modal type of distribution of packet length is obvious.

The autocorrelation coefficients of interarrival times from lag equal to 1 to lag equal to 40 for both total data and without outliers cases are shown in Figures 11 - 16. We can see that data are highly correlated in without outlier cases, and not significantly correlated in the total data case.

Trend analysis, Index of Dispersion Analysis and Comparisons Among Different Measurements. In trend analysis, we try to analyze both total data and without-outliers cases. First, we find the grouping factor k from the autocorrelation coefficient series in measurement 1 and 2. For other measurements, because the autocorrelation coefficient series is never close to zero, we just use the same grouping factor as in measurement 1. The grouping factor found from measurements 1 and 2 is 20, and it is used in all cases. Then, from (3.3) we can find that the theoretical value for the variance of the log chi-squared distribution, v_k, is 0.051269. Our criterion for the best fitting polynomial is the polynomial that has the minimum mean square differences between the origial observed value and the value derived from the fitting polynomial (MSD, in short). But in some cases when the mean square differences for different degrees polynomials are about the same, we will pick the simpler polynomial instead.

The scale unit for the independent variable is either 640 ns (measurements 1 and 2) or 40 ns (all other measurements). The scale unit for the dependent variable is the logarithm of the time units (640 ns or 40 ns).

Index of Dispersion for Interarrival Times. Using the grouping factor k we can calculate the index of dispersion for interarrival times for both cases. Table 2 and 3 show the comparison between the index of dispersion for interarrival times and the square of the coefficient of variation in all measurements. We can find sharp differences between them. It is obvious that none of the measurements can be said to represent a realization of a renewal process.

Table 2. IDI and C^2 comparison (total data case)

Test Case	IDI	C^2
1	0.00063	0.7006
2	0.000632	0.6889
3	0.005281	3.2761
4	0.014826	8.2369
5	0.005293	3.24
6	0.00397	2.0736

Table 3. IDI and C^2 comparison (without outlier case)

Test Case	IDI	C^2
1	0.008985	4.244
2	4.339177	15.288
3	0.061624	71.57
4	0.130347	79.03

Table 3. IDI and C^2 comparison (without outlier case)

Test Case	IDI	C^2
1	0.008985	4.244
2	4.339177	15.288
3	0.061624	71.57
4	0.130347	79.03
5	0.171748	52.85
6	6.819758	1098.92

Measurement 1 (TSOCK Application on SUN4 Host). In this measurement, the traffic source is an application program, TSOCK, on the Sun4 host, let's call it TSOCK1; the traffic destination is another TSOCK on the Sun4 host, let's call it TSOCK2. Data are transmitted from TSOCK1 to TSOCK2, and HILDA sits in between collecting statistics of interarrival times and packet length. TSOCK is running at user priority (lowest priority) and there are heavy system utilizations by other processes, the load average , which is defined as the number of processors needed to process all processes at one time, is about 3.

This measurement and the next are both using a SUN4 host, and thus having a different configuration from other measurements. Here the SUN4 host uses the VME bus to communicate with HILDA, which then interfaces with HiPPI links. Because the VME bus is running only at the rate of 20 Mbps, so even if the HiPPI link is running at 800 Mbps rate, the maximum axevable rate on the HiPPI link can't exceed 20 Mbps. That is, the HiPPI link utilization in this measurement and the next one will be less than 1/40. Indeed, this is what happened as shown in Table 7. The comparison between this traffic pattern, which may be thought as being collected on a link running at the rate of 20 Mbps, and other traffic patterns, which uses the full capacity of HiPPI link, should be interesting and expecting a lot of differences in traffic characteristics.

We have the trend analysis results for both total data and without-outliers cases. Table 8 shows the mean square differences for polynomials of different degrees in both cases, and we have the following fitting results.

1. Total data case:

   The minimum MSD =0.165, is from polynomial with degree equal to 3, and the corresponding coefficients for (3.10) are: C_0 = 2.344133, C_1 = 0.222412, C_2 = -0.007596 and C_3 = -0.020430.

2. Without-outliers data:

   The minimum MSD=0.014, is from polynomial with degree equal to 1, and the corresponding coefficients for (3.10) are: C_0 = 5.000204, and C_1 = 0.000698 .

It is interesting to see that in without-outliers case C_1 =0.000698; it is almost like no trend existed at all. That is, we can treat it as a stationary process analysis. But when the outliers are included, these outliers serve as "disturbance factors", and the process is no longer stationary.

The stationarity of the without-outliers case is not seen in other measurements. One possible explanation is that because the VME bus makes the effective HiPPI link rate down to 20 Mbps, and the traffic characteristics of very high speed network isn't shown in this case.

Measurement 2 (TSOCK Application on SUN4 Host with Induced Error Bits). This measurement has the same configuration as measurement 1. The only difference is that, in order to see how erroneous data affect traffic pattern, we have HILDA induce error bits into the data stream; one erroneous bit is induced for every 10^8 bits transmitted. This introduction of erroneous bits causes TCP/IP error recovery scheme to be activated.

We have the trend analysis results for both total data and without-outliers cases. Table 9 shows the mean square differences for polynomials of different degrees in both cases, and we have the following fitting results.

1. Total data case:

The minimum MSD=0.171, is from polynomial with degree equal to 1, and the corresponding coefficients are: C_0 = 5.129234 and C_1 = 0.130373.

2. Without-outliers case:

Although the minimum MSD is from a polynomial with degree equal to 8, the polynomial of degree one which has the MSD= 0.033 is also a fairly good estimate. For simplicity, we will pick the polynomial of degree one, and the corresponding coefficients are: C_0 = 5.099341 and C_1 = 0.107016.

We can see that the traffic patterns are very similar in both cases. When comparing this measurement with measurement 1 for without-outliers cases, we can see the induced error bits cause the originally almost-stationary process into a non-stationary process.

Measurement 3 (FTP Application on Cray Host). In this measurement, the traffic source is an application program FTP on the Cray-YMP supercomputer, let us call it FTP1; The traffic destination is another FTP on a Convex host, call it FTP2. In between sits an Ultra network as a gateway of two heterogeneous networks. HILDA is collecting the traffic on HIPPI link. At first glance, it seems that the Ultra network and the Convex host have nothing to do with the traffic data being collected. They actually affect the traffic pattern indirectly. Because the Ultra network and the Convex host can't process data as fast as Cray-YMP, when Cray-YMP pumps in more data than they can handle, their flow control mechanisms will be invoked, and the data traffic will be slowed down.

The traffic characteristics in this measurement differs from measurement 1 greatly. We can see the mean is about $\frac{1}{4}$ of that of measurement 1 and the coefficient of variation is about four times of that of measurement 1 (see Table 4). That is, the traffic are far more bursty than in measurement 1. There are two major reasons. First, Cray-YMP, with its processing power, can pump data into transmission much faster than the SUN4 station, this makes the interarrival times shorter and traffic more bursty. Second, unlike in measurement 1, the full capacity of the HiPPI link has been used; with different link capacity, we have different traffic patterns. We can also see an extremely long tail of outliers, much longer than in measurement 1. The link utilizations are also much higher than in measurement 1.

We have the trend analysis results for both total data and without-outliers cases. Table 10 shows the mean square differences for polynomials of different degrees in both cases, and we have the following fitting results.

1. Total data case:

The minimum MSD=0.256, is from polynomial with degree equal to 2, and the corresponding coefficients are: C_0 = 3.493535, C_1 = 0.189942, and C_2 = 0.075997.

2. Without-outliers case:

The minimum MSD=0.136, is from polynomial with degree equal to 1, and the corresponding coefficients are: C_0 = 3.432610, and C_1 = 0.325506.

Measurement 4 (UFTP Application on Cray Host). In this measurement, the traffic source is an application program UFTP on the Cray-YMP supercomputer. The traffic destination is an FTP on the Convex host. The network configuration is the same as in measurement 3. We would like to know if the traffic characteristics has been changed because of the different application program being used. This measurement and the next measurement from TSOCK will be used to compare with the previous measurement from FTP.

The mean packet length is about 7 times of the previous measurement (see Table 6). This means that less packet overhead is used in the data transmission. The mean interarrival times for without-outlier case is about $\frac{1}{3.5}$ of that of the total data case. The outliers' effect on mean seems unusually large (see Table 4). The reason is that we have three unusually large outliers, on average about $3.10 * 10^7$ (40ns time unit), at the beginning of the measurement, and have the last outlier greater than $15 * 10^7$. Even with these four extreme large outliers, the HiPPI link utilizations are 5.3% and 19% respectively for the total data case and for the without-outliers case. This is far better than the previous measurement (see Table 7). There are three possible reasons. First, UFTP interfaces with HiPPI interface directly, so less transmission overhead is used, and data get transmitted faster. Thus, better link utilization is obtained. Second, UFTP is not a standard like FTP, it is used only in test environments. In most cases, it becomes a dedicated file server, and can process file transfer much faster than FTP. Third, longer packet length makes transmission more efficient.

We have the trend analysis results for both total data and without-outliers cases. Table 11 shows the mean square differences for polynomials of different degrees in both cases, and we have the following fitting results.

1. Total data case:

   The minimum MSD=0.732, is from polynomial with degree equal to 6, and the corresponding coefficients are: C_0 = 4.049303, C_1 = 0.473343, C_2 = -1.295650, C_3 = 1.809494, C_4 = -0.997018, C_5 = 0.677519, and C_6 = -0.405313.

2. Without-outliers data:

   The minimum MSD=0.276, is from polynomial with degree equal to 2, and the corresponding coefficients are: C_0 = 3.403847, C_1 = 0.569093, and C_2 = -0.211580.

The degree of the fitting polynomial is high and MSD is very large in the total data case; that is, the fitting result is not as good as in other measurements. The reasonable guess is that in this measurement we have four unusual large outliers.

Measurement 5 (TSOCK Application on Cray Host). In this measurement, the traffic source is an application program TSOCK on the Cray-YMP supercomputer and the traffic destination is another TSOCK on the Convex host. The network configuration is the same as measurement 3.

The traffic characteristics is similar to the results obtained from measurement 3. This is not a surprise, because TSOCK, like FTP, also uses TCP/IP to transmit data. However, the packet length is a little bit longer than that obtained in measurement 3. The HiPPI link utilizations are a little bit higher than those from measurement 3. This may be coming from the fact that TSOCK, like UFTP, is not a standard, and is used only in test environments. In most cases, it becomes a dedicated application program, and can process file transfer faster than FTP.

We have the trend analysis results for both total data and without-outliers cases. Table 12 shows the mean square differences for polynomials of different degrees in both cases, and we have the following fitting results.

1. Total data case:

   The minimum MSD=0.423, is from polynomial with degree equal to 1, and the corresponding coefficients are: C_0 = 3.511850 and C_1 = -0.021990.

2. Without-outliers data:

   The minimum MSD=0.109, is from polynomial with degree equal to 1, and the corresponding coefficients are: C_0 = 3.435505 and C_1 = 0.138831 .

Measurement 6 (UFTP Application on Cray Host with 4 Concurrent File Transfers). In this measurement, the traffic source is an application program UFTP on Cray-YMP. There are four ports using UFTP for file transfer, and these file transfers are concurrent. The aggregate traffic data are collected by HILDA. The traffic destination is another FTP on the Convex host. The network configuration is the same as in measurement 3.

Since we have four concurrent data transmitting streams, we are going to have cross-correlation on our collected data. It will be worth investigating how much the impact from cross-correlation is. We can see that data are highly correlated in without-outlier case, and not correlated at all in the total data case. As comparing with other measurements, the coefficients of autocorrelation do fluctuate more than cases 3 , 4 and 5 (see Figure 16). The fluctuations are bigger when the lags are small and are dying down as the lags become larger. This kind of fluctuation is not seen in other measurements, the possible guess is that it reflects the cross-correlation effects.
Also noticeable, we have an extremely large coefficient of variation in total data case, and a very small coefficient of variation in the without-outliers case (see Table 5).

The reason for the extremely large coefficient of variation is because we have a cluster of extremely large outliers at the beginning of the measurement. These outliers are related with waiting for the Cray-YMP's preemption. Apparently UFTP is not an active process running on Cray-YMP at the beginning of the measurement, it has to wait for the Cray-YMP's preemption so that it can get swapped in. The waiting time contributes a cluster of extremely large outliers, thus we have the largest coefficient of variation.

There is one possible reason that we have the smallest coefficient of variation in all without-outliers cases. Because we have four concurrent data transmissions, there are fewer "big gaps" in the total transmission period, that is the discrepancies among interarrival times are not as large as other measurements; and this contributes a smaller coefficient of variations when those outliers are excluded. Also this fewer "big gaps" effects in the transmission period makes the mean interarrival times of this measurement shortest among all measurements.

We have the trend analysis results for both total data and without-outliers cases. Table 13 shows the mean square differences for polynomials of different degrees in both cases, and we have the following fitting results.

1. Total data case:

   The minimum MSD=0.255, is from polynomial with degree equal to 5, and the corresponding coefficients are: C_0 = 2.027890, C_1 = 1.721929, C_2 = -0.930602, C_3 = -0.199233, C_4 = 0.413372, and C_5 = -0.362120.

2. Without-outliers case:

   The minimum MSD is from polynomial with degree equal to 9, but the polynomial with degree 1 with MSD=0.083 is pretty good too. Thus we pick degree one and the corresponding coefficients are: C_0 = 2.991391, and C_1 = 0.356826.

Summary of the Trend Analyses for Interarrival Times. Two conclusions can be drawn from the above trend analysis results. First, the fitting results from the total data cases are not really good. Their MSD values are higher, and sometimes the degree of fitting polynomials is very high. If we look at their MSD distributions for different degree of polynomials, we may find out they are extremely unstable. So, even when we can find a fitting polynomial with reasonable MSD, its reliability is still in question. Because the measurement data for total data cases are not highly correlated, we don't expect that any trend can be found for un-correlated measurement data. Second, the fitting results from without-outliers cases are very good. Their MSD values are smaller and reasonably close to the theoretical value for the variance of the log chi-squared distribution. The degree of fitting polynomials is simple, we only have one measurement that needs polynomial of degree 2 to fit, in all other measurements, polynomials with degree one can fit fairly well. If we look at the MSD distributions for different degree of polynomials, we may find out they are very stable. This kind of results are just as expected, because the measurement data for without-outlier cases are highly correlated, and we can expect there are trends in the correlated measurement data. More important is that these without-outliers measurements represent the data transmission without the interference of background environment.

Table 4. Mean interarrival times comparison. Time unit is ms.

Measurement Number	1	2	3	4	5	6
Total Data Case	10	10.5	2.41	5.6	2.98	1.7
Without Outliers Case	7.2	7.3	1.88	1.6	1.86	1.02

Table 5. Coefficient of variation of interarrival times comparison

Measurement Number	1	2	3	4	5	6
Total Data Case	2.06	3.91	8.46	8.88	7.27	33.2
Without Outliers Case	0.84	0.83	1.81	2.87	1.8	1.44

Table 6. Mean packet length comparison. Length unit is word.

Measurement Number	1	2	3	4	5	6
Total Data Case	1019	1019	1024	7441	1653	2112
Without Outliers Case	1024	1024	1027	7614	1662	2114

Table 7. HiPPI link utilization (in %) comparison

Measurement Number	1	2	3	4	5	6
Total Data Case	0.41	0.39	1.7	5.3	2.2	5.0
Without Outliers Case	0.57	0.56	2.2	1.9	3.6	8.3

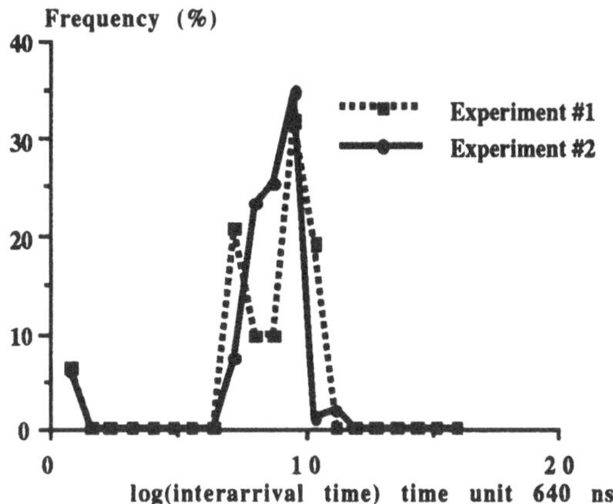

Figure 5. Interarrival time distribution of measurements 1 and 2.

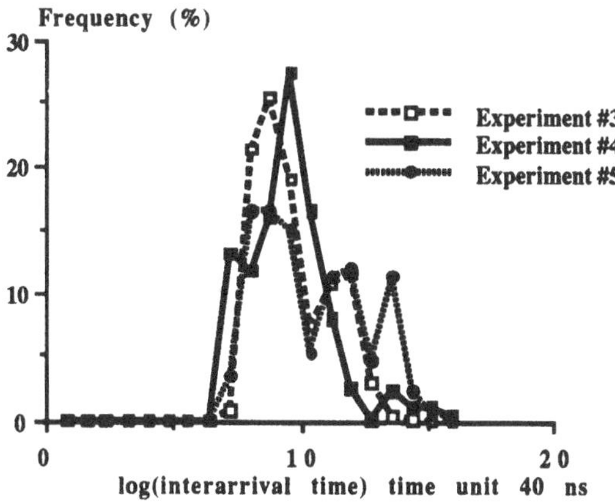

Figure 6. Interarrival time distribution of measurements 3, 4 and 5.

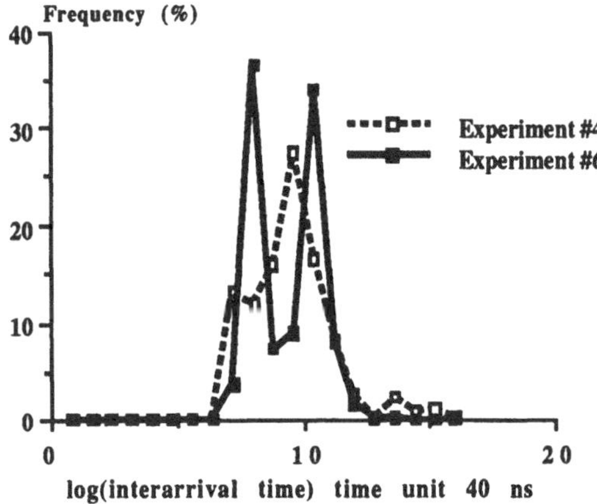

Figure 7. Interarrival time distribution of measurements 4 and 6.

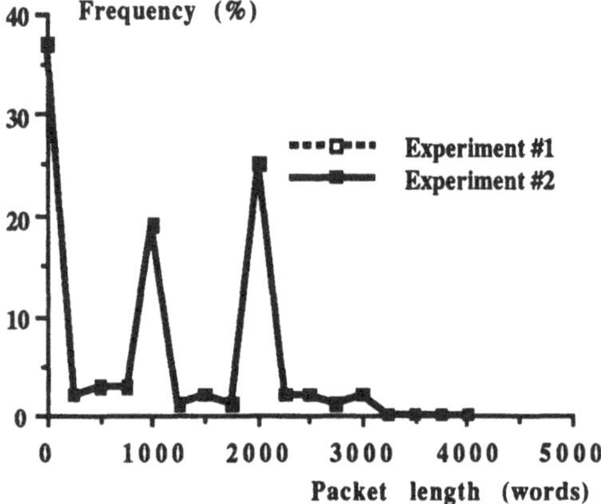

Figure 8. Packet length distribution of measurements 1 and 2.

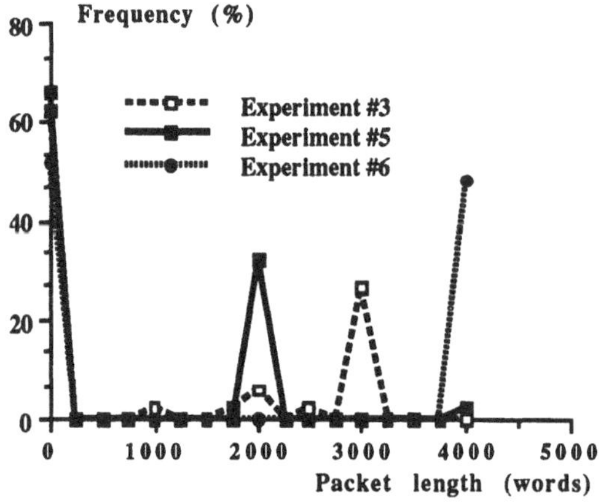

Figure 9. Packet length distribution of measurements 3, 5 and 6.

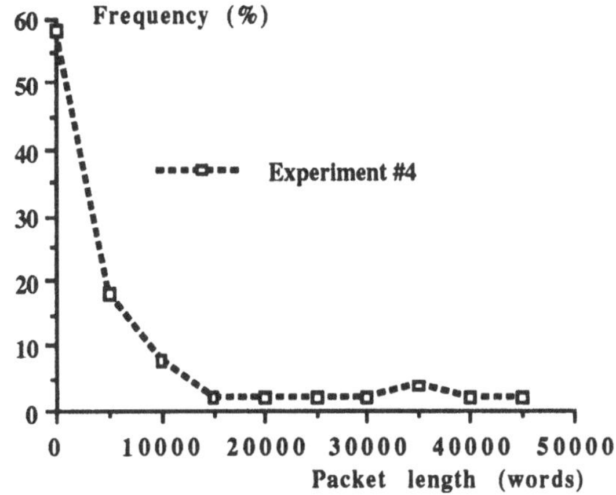

Figure 10. Packet length distribution of measurement 4.

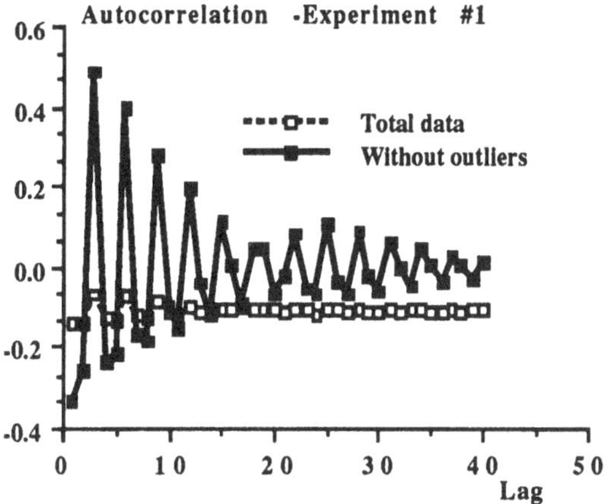

Figure 11. Autocorrelation coefficient distribution of measurement 1.

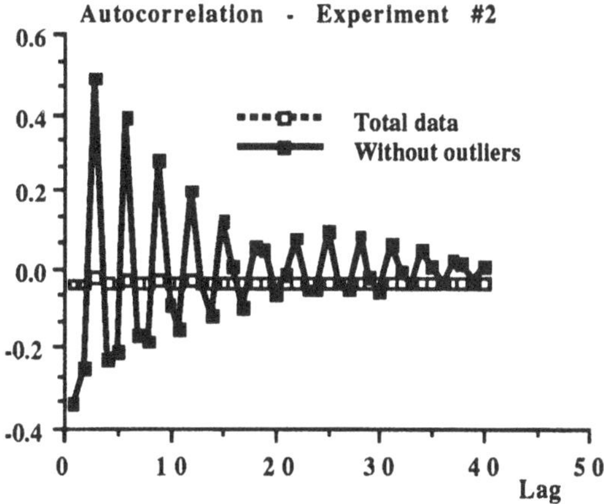

Figure 12. Autocorrelation coefficient distribution of measurement 2.

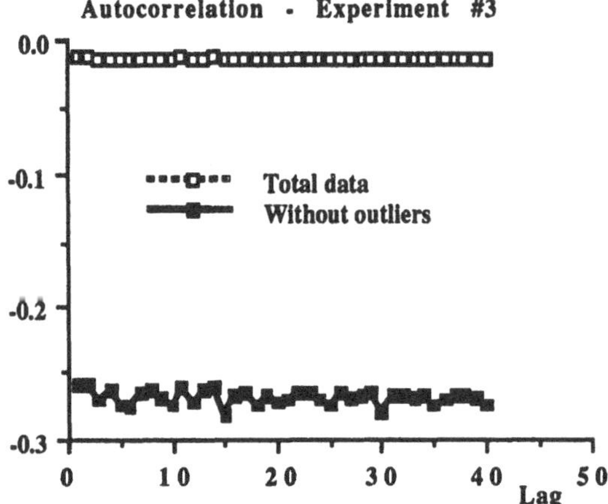

Figure 13. Autocorrelation coefficient of measurement 3.

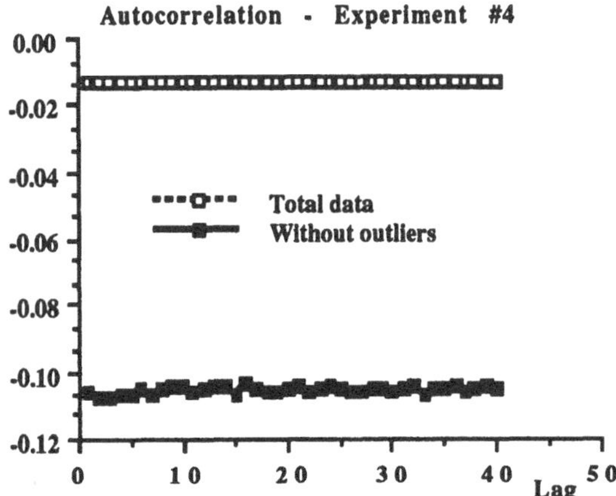

Figure 14. Autocorrelation coefficient distribution of measurement 4.

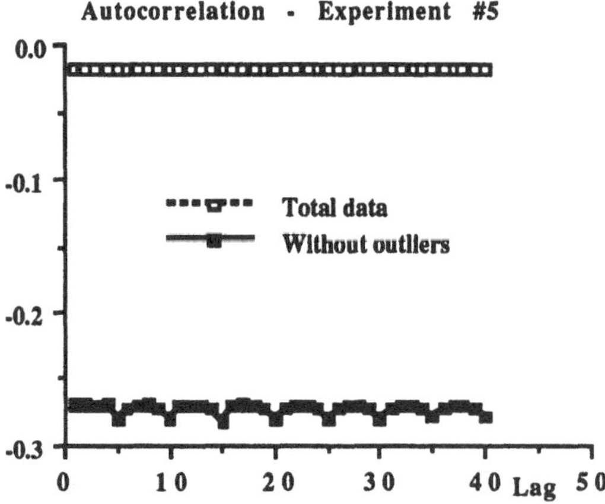

Figure 15. Autocorrelation coefficient distribution of measurement 5.

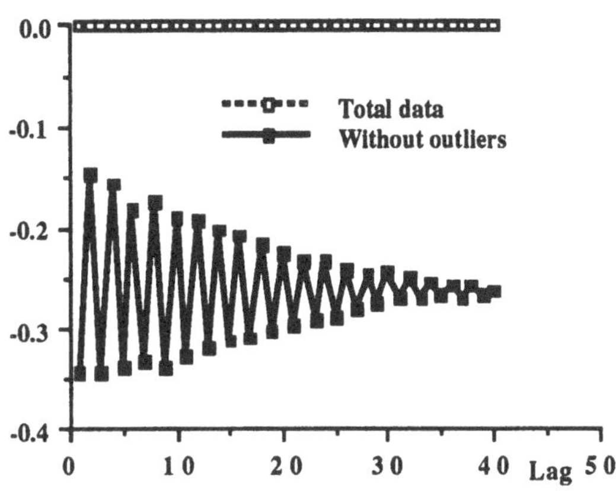

Figure 16. Autocorrelation coefficient distribution of measurement 6.

Table 8. Mean square differences (MSD) for different degrees of polynomial fitting (Measurement 1). TSOCK on SUN4.

Degree of Polynomial	1	2	3	4	5	6	7	8	9	10
MSD (total data case)	0.413	0.615	0.165	0.365	0.314	0.629	0.179	0.328	0.229	0.306
MSD (without outliers case)	0.014	0.015	0.028	0.017	0.027	0.016	0.017	0.030	0.022	0.025

Table 9. Mean square differences (MSD) for different degrees of polynomial fitting (Measurement 2). TSOCK on SUN4 with error bits induced.

Degree of Polynomial	1	2	3	4	5	6	7	8	9	10
MSD (total data case)	0.171	0.218	0.404	0.266	0.500	0.269	0.237	0.749	0.516	0.520
MSD (without outliers case)	0.033	0.019	0.021	0.024	0.025	0.022	0.017	0.018	0.047	0.024

Table 10. Mean square differences (MSD) for different degrees of polynomial fitting (Measurement 3). FTP on Cray-YMP.

Degree of Polynomial	1	2	3	4	5	6	7	8	9	10
MSD (total data case)	0.264	0.256	1.218	0541	28200	0.617	1.342	346.8	83.13	4105
MSD (without outliers case)	0.136	0.114	0.275	0.156	0.101	0.149	0.116	0.189	0.118	0.100

Table 11. Mean square differences (MSD) for different degrees of polynomial fitting (Measurement 1). UFTP on Cray-YMP with 1 port.

Degree of Polynomial	1	2	3	4	5	6	7	8	9	10
MSD (total data case)	0.884	7.670	0.854	67358	0.966	3.433	3.520	1.484	1.380	0.165
MSD (without outliers case)	0.413	0.276	0.389	0.836	0.303	0.324	0.501	0.340	0.616	0.461

Table 12. Mean square differences (MSD) for different degrees of polynomial fitting (Measurement 1). TSOCK on Cray-YMP.

Degree of Polynomial	1	2	3	4	5	6	7	8	9	10
MSD (total data case)	0.423	1.232	5.764	0.510	0.488	1032	218.9	4.234	$1.1 * 10^7$	1.954
MSD (without outliers case)	0.109	0.150	0.094	0.177	0.129	0.103	0.147	0.102	0.197	0.123

Table 13. Mean square differences (MSD) for different degrees of polynomial fitting (Measurement 1). UFTP on Cray YMP with 4 ports.

Degree of Polynomial	1	2	3	4	5	6	7	8	9	10
MSD (total data case)	1.355	0.380	1.934	0.507	$3.5 * 10^{11}$	0.328	0.599	0.445	14.80	0.306
MSD (without outliers case)	0.083	0.098	0.091	0.130	0.091	0.090	0.089	0.113	0.081	0.104

REFERENCES

[BA46] M. Bartlett, "On the theoretical specification and sampling properties of autocorrelated time series," J.R. Statist. Soc. (Suppl,), 8, 27-41, 1946.

[BH90] American National Standard for information Systems, Broadband HiPPI/SONET Terminal Adapter Standards Report, (BHS-TA), Rev 1.0 July, 6, 1990.

[BO91] L. Bottomley, A. Nilsson, and A. Blateky, "Traffic measurements on a working wide area network," ITC, Copenhagen, June 1991.

[CO66] D. Cox, and P. Lewis, The Statistical Analysis of Series of Events, John Wiley and Sons, Inc., 1966.

[DA33] H. Davis, Table of the Higher Mathematical Functions, Vol 1, Bloomington, Indiana: Principia Press, 1933.

[FO91] H. Fowler and W. Leland, "Local area network traffic characteristics, with implications for broadband network congestion management," IEEE J-SAC, Vol. 9, No. 7, pp 1139-1149, Sep. 1991.

[GU91] R. Gusella, "Characterizing the variability of arrival processes with indexes of dispersion," IEEE J-SAC, Vol. 9, No. 2, pp 203-211, Feb. 1991.

[HL90] American National Standard for information Systems, HiPPI Link Encapsulation Layer(HiPPI-LE), X3T9.3/90-119, 1990.

[JA86] R. Jain and S. Routhier, "Packet trains-measurements and a new model for computer network traffic," IEEE J-SAC, Vol. 4, No. 6, pp 986-995, Sep. 1986.

[KU71] S. Kuo, Computer Applications of Numerical Methods, Addison-Wesley Inc., 1971.

[LA90] A. Lazar, G. Pacific, and J. White, "Real time traffic measurements on MAGNET II," IEEE J-SAC, Vol. 8, No. 3, pp 467-483, Apr. 1990.

[NO89] M. Nomura, T. Fujii, and N. Ohta, "Basic characteristics of variable rate coding in ATM environment," IEEE J-SAC, Vol. 7, No. 5, pp 752-760, June 1989.

[WI92] D. Winkelstein and D. Stevenson "HiPPI Link Analysis System test equipment for high speed network analysis," MCNC Communication Dept. NC, 1992.

INDEX

AAL, 77
AAL-3/4, 17, 18
AAL-5, 17, 18
Access classes, 44
Active carrier, 10
Address screening, 44
Address validation, 43
Aging factor, 167-175
Application specific integrated circuits, 41
ARMA, 183-184
ATM, 3, 5, 6, 8-12, 15-17, 19-26, 31-32, 34-38, 42, 89, 91, 100, 101, 103, 104, 107, 108, 116, 117, 123, 124, 131, 135, 157, 180-182, 185
ATM adaptation layer, 11, 22, 180
ATM adapter, 6
ATM cells, 16, 18, 24
ATM forum, 21-23, 27
ATM layer, 22, 24
ATM switch, 6, 7, 12
Autocorrelation coefficients, 206, 210-212, 214, 218
Autocorrelation function, 182-184
Autocovariance function, 210

B-pictures, 181-182
Bandwidth allocation, 17
Bandwidth enforcement, 135, 158
Bandwidth management, 17
BERKOM project, 79
Bernoulli process, 158
Binary Markov sources, 123, 129, 130
BISDN, 21, 107-108, 116-117, 157, 179, 185, 199
Block matching algorithm, 182
Broadband networks, 16, 18, 39
Buffer management, 157-158
Burst tolerance, 26
Burstiness, 78, 111, 113
Burstiness coefficient, 142
Bus arbitration, 35

Caching, 36
Call admission control, 157
Call setup, 41
Call teardown, 41
CBR, 180
CCITT, 11, 17, 21-25, 27, 45, 107, 110, 179, 191
Cell discard, 23, 25, 157
Cell jitter, 113
Cell loss priority, 24, 26-27, 182-183
Cell loss probability, 109, 123, 124, 136, 144, 146, 157-159, 180, 199
Cell loss rate, 157-158, 160-162
Cell ordering, 19
Cell payload, 16
Cell relay, 21, 23
Cell spacing algorithm, 111
CENTREX, 12
Chebyshev polynomials, 207-208
Choke packet, 116
Circuit emulation, 50, 58, 68
Circuit oriented services, 19
Circuit switching, 3, 7, 9
Client server model, 33
CLNP protocol, 79
Coding delay, 181
Collection services, 192
Committed information rate, 46
Common format, 45
Congestion control, 157, 196
Connection acceptance algorithm, 158
Connection admission control, 23-24, 108-110, 114, 117, 157, 159
Connection oriented transport, 41, 79, 157
Connectionless service, 3, 79
Convergence layer, 77
Conversational services, 180, 190-191, 193
CRAY YMP, 203, 212-213, 216, 218
Credit based congestion control, 150

Data corruption, 82
Data exchange interface, 22-23
Data manipulation, 84
Data replication, 82
Data transfer, 84
Decoding delay, 181
Dial tone delay, 167
Digamma function, 204
Directory service, 8, 10-11
Discrete cosine transform, 182

Dispersion index, 210, 213
Distributed source control, 112
Distribution services, 179, 193
DLC, 32
DLCI, 40
DQDB, 39, 55-56, 63-68, 71
DS-0 signal, 51
DS-1 signal, 42, 51, 53, 169
DS-3 signal, 8, 44, 51, 59
DSU, 43
DXI, 43, 47
Dynamic bandwidth capability, 46
Dynamic time windows, 117

Echo, 16
Echo cancellation, 193
Effective bus bandwidth, 35
Elastic buffer, 32
Enhanced transport service interface, 78
Equivalent bandwidth, 110
ESPRIT project, 75
Ethernet, 6, 31, 34
Explicit congestion notification, 45, 116, 117
Explicit forward congestion notification, 116
Exponentially weighted moving average, 111, 112, 136

Fast packet switching, 15, 16
Fast resource allocation, 114, 115
FDDI, 4, 6, 8, 31, 34, 39, 59, 89, 91-92, 100-102, 104
FDDI-II, 8
Fiber channel standard, 3-5, 7-9
FIFO service, 66
First order Markov process, 142
Flexible communication subsystem, 29
Fluid flow model, 110, 124
Flush bit, 36
Four wire circuit, 16
FPDL, 83
Fragmentation, 17
Frame relay, 9, 22, 39, 49, 51-52, 54
Frame relay header, 43
FTP, 203, 216-217

Gamma function, 204-205
Global addressing mode, 45
Grade of Service, 123, 125, 131, 137, 139, 167-170, 172, 175-176
Group addressing, 44
Group interprocess communication, 90

HDTV, 50, 181, 189, 193
High speed networks, 199-200, 202, 210, 212-213
HILDA, 200-202, 215-217
HiPPI, 2, 3, 17-18, 49-50, 200-204, 210-217
HiPPI host mode, 202
HiPPI-FP, 200, 203
HiPPI-IPI, 200
HiPPI-LE, 203
HiPPI-MP, 200
HiPPI-PH, 200-201, 203
HOP structure, 80
HSLAN, 75

I-pictures, 181-182
IBP, 204
Implicit congestion notification, 45
Integrated video services baseline, 179
Intelligent peripheral interface, 3
Intercarrier interface, 22-23
Interdata service, 47
Interexchange carriers, 47
Interleaving, 182
IP, 90, 92
IPI3, 3
IPP, 204
ISDN, 9, 19, 41
Isochronous traffic, 3, 8-9, 67-68

Jitter, 78, 192-196
JTC1/SC6, 76
Jumping window, 111-113, 116, 136

Keep alive sequence, 45

L3 PDU, 43
LAN, 2, 4-6, 8, 11, 17, 18, 22, 34, 194, 196, 199, 210
LAN interconnection, 39
Latency, 190-196
Leaky bucket, 17, 25, 27, 110-113, 116, 135-140, 158, 180
Least squares, 206, 208
Local ATM network, 10
Local management interface, 45
Log chi squared distribution, 204, 206, 214, 218
Logical private networks, 44
Long distance network, 16, 19

$M|G|m$ approximation, 172, 175-176
MAC, 32, 77, 90, 92, 195
MAN, 8, 11, 76
Management control, 10
Markov modulated Bernoulli process, 108
Maximum transit delay, 46
MCNC, 203
Messaging services, 180, 192
Metasignalling, 59
MMPP, 204, 210
Modular communication machine, 80
Modulating Markov chain, 125
Moving window, 111, 112, 136
MPEG, 181-182, 185
Multicast, 45, 89-93, 95-97, 101, 103-104
Multimedia, 1, 12
Multimedia applications, 3, 6, 58, 75, 187, 193, 196
Multimedia servers, 6
Multimedia services, 49
Multimedia traffic, 157
Multipoint calls, 180

Network data collection, 42
Network terminal adapter, 18
Node to node interface, 42
Nonblocking switch, 2
NTSC channels, 19

OAM, 45
Operation and maintenance cells, 26
Optical level, 168
Optoelectronic converters, 7
Orthogonal polynomials, 206-209
Outliers, 212

P-pictures, 181-182
Packet interleaving, 32, 34
Packet replication, 90
Packet segmentation and reassembly, 31
Packet switching, 2, 7
Partial buffer sharing, 114, 158-162, 164
PBX, 5, 12
Peak bandwidth, 17
Peak rate, 23, 25-27, 109, 110, 117, 123, 130, 136, 137
Permanent virtual circuit, 23, 40
Permanent virtual circuit status, 45
Point to multipoint signalling, 22
Pointer prefetching, 36
Poisson process, 204, 210
Policing mechanism, 17, 110-114, 117
POTS, 69
Predefined service classes, 82
Preventive congestion control, 107-108, 115-117
Prioritized access, 168
Priority control, 23, 114
Private internodal protocols, 10
Private UNI, 26
Protocol configurator, 83
Protocol data unit, 42
Provisioning, 11
Public UNI, 26
Pulse scheduling, 116
Pushout, 114, 158

Q.931, 17
Q.93B specification, 22
QPSX, 64
Quality of service, 23-24, 78, 108, 143, 179

Rate control, 84
RBOC, 45, 56
Reactive control, 107-108, 115-117
Real time traffic, 158
Reassembly, 17
Regression analysis, 205-206
Relaxed cell loss ratio, 181
Relief, 141
Relief function, 135
Relief mechanism, 136, 138-140
Renewal process, 210, 214
Residual error rate, 78
Resource management tools, 83
Restoration, 42
Retrieval services, 180, 192
Route computation, 8, 41
Route establishment, 10-11
Router, 9-10
Router DSU, 43
Router vendor, 43
Routing table dissemination, 90

SAR, 32, 43, 77
SDH, 59
Service availability, 46
Session manager, 83
Sigma relief, 145
SIP, 44
Skew, 192-193, 196
Sliding window, 112, 115
SMDS, 9, 22, 23, 39, 42, 49, 52-54, 56
Smoothing buffer, 196
SONET, 8, 11, 18-19, 50, 59, 168-169
SONET OC-3, 8, 11, 18, 34
SONET OC-12, 8, 18
Source address, 44
Space priority, 157-158, 160
Spacing algorithm, 136
Spatial redundancy, 182
Staging buffer, 20
Stop and go protocol, 112, 113
Striping, 19
STS-3, 19, 169
STS-3c, 42, 59
Subband video encoder, 158
Switch-based LAN, 4-6
Switched virtual circuit, 11, 41
Synchronizing handshake, 92
Synchronous service, 76

T-carrier transmission systems, 49
T1 link, 9, 19, 42, 181
T3 link, 9, 19, 34, 42
Tagging, 114, 117
TCP/IP, 16, 77, 203, 213, 215, 217
Ternary Markov source, 129, 130
Threshold pushout, 159, 162, 164
Token ring, 6, 31, 34
TP4, 79
Traffic contract, 23, 24, 26
Traffic descriptor, 23, 26, 27, 69
Traffic management, 22-23, 25, 27, 45
Traffic regulation, 135, 136, 138, 139
Traffic shaping, 27, 113, 117
Transport layer, 46
Transport protocol data unit, 89-91
Transport service, 75
Trunk line, 42
TSOCK, 203, 215-217
Two-wire circuit, 16

UDP, 78

UFTP, 203, 217
UNI, 22-23, 26, 112-113, 117
Unicast, 90, 103
Unshielded twisted pair, 7-8
Usage parameter control, 23-24, 27, 110, 180-183

V.35, 23
Variable bit rate traffic, 199
Variable size packets, 31-32, 38
Variation coefficient, 212, 214, 218
VBR video, 180, 199
Violation tag, 113, 117
Virtual bandwidth, 108, 110
Virtual circuit, 41, 124, 157, 180-182
Virtual clock, 113, 116
Virtual path, 124
Virtual tributary, 169-170, 172, 175-176
Vistanet, 16, 18
VME bus, 213, 215
VME host, 202
VMTP, 90
Voice annotate text, 189
Voice network, 12, 16, 18

WAN, 8-11, 17, 22, 34, 76, 199, 210
Window flow control, 112, 115-117
Window policing, 111
Wiring cabinet, 4

X-kernel approach, 80
X.25 packets, 51
X3T9.3 standard, 200
XON/XOFF mechanism, 45
Xpress transfer protocol, see XTP
XTP, 89-100, 103-104

Zeitfracht Medien GmbH
Ferdinand-Jühlke-Straße 7
99095 Erfurt, Deutschland
produktsicherheit@kolibri360.de